IFIP Advances in Information and Communication Technology

483

Editor-in-Chief

Kai Rannenberg, Goethe University Frankfurt, Germany

Editorial Board

IFIP – The International Federation for Information Processing

IFIP was founded in 1960 under the auspices of UNESCO, following the first World Computer Congress held in Paris the previous year. A federation for societies working in information processing, IFIP's aim is two-fold: to support information processing in the countries of its members and to encourage technology transfer to developing nations. As its mission statement clearly states:

IFIP is the global non-profit federation of societies of ICT professionals that aims at achieving a worldwide professional and socially responsible development and application of information and communication technologies.

IFIP is a non-profit-making organization, run almost solely by 2500 volunteers. It operates through a number of technical committees and working groups, which organize events and publications. IFIP's events range from large international open conferences to working conferences and local seminars.

The flagship event is the IFIP World Computer Congress, at which both invited and contributed papers are presented. Contributed papers are rigorously refereed and the rejection rate is high.

As with the Congress, participation in the open conferences is open to all and papers may be invited or submitted. Again, submitted papers are stringently refereed.

The working conferences are structured differently. They are usually run by a working group and attendance is generally smaller and occasionally by invitation only. Their purpose is to create an atmosphere conducive to innovation and development. Refereeing is also rigorous and papers are subjected to extensive group discussion.

Publications arising from IFIP events vary. The papers presented at the IFIP World Computer Congress and at open conferences are published as conference proceedings, while the results of the working conferences are often published as collections of selected and edited papers.

IFIP distinguishes three types of institutional membership: Country Representative Members, Members at Large, and Associate Members. The type of organization that can apply for membership is a wide variety and includes national or international societies of individual computer scientists/ICT professionals, associations or federations of such societies, government institutions/government related organizations, national or international research institutes or consortia, universities, academies of sciences, companies, national or international associations or federations of companies.

More information about this series at http://www.springer.com/series/6102

Youngsoo Shin · Chi Ying Tsui
Jae-Joon Kim · Kiyoung Choi
Ricardo Reis (Eds.)

VLSI-SoC: Design for Reliability, Security, and Low Power

23rd IFIP WG 10.5/IEEE International Conference
on Very Large Scale Integration, VLSI-SoC 2015
Daejeon, Korea, October 5–7, 2015
Revised Selected Papers

 Springer

Editors
Youngsoo Shin
KAIST
Daejeon
Korea (Republic of)

Chi Ying Tsui
Hong Kong University of Science and
  Technology
Clear Water Bay
Hong Kong

Jae-Joon Kim
POSTECH
Pohang
Korea (Republic of)

Kiyoung Choi
Seoul National University
Seoul
Korea (Republic of)

Ricardo Reis
Federal University of Rio Grande do Sul
Porto Alegre, Rio Grande do Sul
Brazil

ISSN 1868-4238   ISSN 1868-422X (electronic)
IFIP Advances in Information and Communication Technology
ISBN 978-3-319-83440-5   ISBN 978-3-319-46097-0 (eBook)
DOI 10.1007/978-3-319-46097-0

Printed on acid-free paper

This Springer imprint is published by Springer Nature
The registered company is Springer International Publishing AG Switzerland

Preface

This book contains extended and revised versions of the highest quality papers, which were presented during the 23rd IFIP/IEEE WG10.5 International Conference on Very Large Scale Integration (VLSI-SoC), a global System-on-Chip Design & CAD conference. The 23rd conference was held at the Daejeon Convention Center, South Korea, during October 5–7, 2015. Previous conferences have taken place in Edinburgh, Scotland (1981); Trondheim, Norway (1983); Tokyo, Japan (1985); Vancouver, Canada (1987); Munich, Germany (1989); Edinburgh, Scotland (1991); Grenoble, France (1993); Chiba, Japan (1995); Gramado, Brazil (1997); Lisbon, Portugal (1997); Montpellier, France (2001); Darmstadt, Germany (2003); Perth, Australia (2005); Nice, France (2006); Atlanta, USA (2007); Rhodes Island, Greece (2008); Florianopolis, Brazil (2009); Madrid, Spain (2010); Kowloon, Hong Kong (2011), Santa Cruz, USA (2012), Istanbul, Turkey (2013), and Playa del Carmen, Mexico (2014).

The purpose of this conference, which was sponsored by IFIP TC 10 Working Group 10.5, the IEEE Council on Electronic Design Automation (CEDA), and by IEEE Circuits and Systems Society, with the In-Cooperation of ACM SIGDA, was to provide a forum for the exchange of ideas and presentation of industrial and academic research results in the field of microelectronics design. The current trend toward increasing chip integration and technology process advancements has brought new challenges both at the physical and system design levels, as well as in the test of these systems. VLSI-SoC conferences aim to address these exciting new issues.

The quality of submissions (117 regular papers from 28 countries, excluding PhD Forum and special sessions) made the selection process a very difficult one. Finally, 44 submissions were accepted as full papers and 17 as posters. Out of the 44 full papers presented at the conference, 10 papers were chosen by a selection committee to have an extended and revised version included in this book. The selection process of these papers considered the evaluation scores during the review process as well as the review forms provided by members of the Technical Program Committee and Session Chairs as a result of the presentations.

The chapters of this book have authors from China, Denmark, France, Germany, Hong Kong, Italy, Ireland, South Korea, The Netherlands, Switzerland, and the USA. The Technical Program Committee comprised 92 members from 24 countries.

VLSI-SoC 2015 was the culmination of the work of many dedicated volunteers: paper authors, reviewers, session chairs, invited speakers, and various committee chairs. We thank them all for their contribution.

This book is intended for the VLSI community, mainly those persons who did not have the chance to attend the conference. We hope you will enjoy reading this book and that you will find it useful in your professional life and for the development of the VLSI community as a whole.

August 2016

Youngsoo Shin
Chi Ying Tsui
Jae-Joon Kim
Kiyoung Choi
Ricardo Reis

Organization

The IFIP/IEEE International Conference on Very Large Scale Integration-System-on-Chip (VLSI-SoC) 2015 took place during October 5–7, 2015 in the Daejeon Convention Center, South Korea. VLSI-SoC 2015 was the 23rd in a series of international conferences, sponsored by IFIP TC 10 Working Group 10.5 (VLSI), IEEE CEDA, and ACM SIGDA. The organization of the conference was done by the following people:

General Chairs

Naehyuck Chang	KAIST, South Korea
Kiyoung Choi	Seoul National University, South Korea

Technical Program Chairs

Youngsoo Shin	KAIST, South Korea
Chi-Ying Tsui	HKUST, Hong Kong, China

Technical Vice Program Chair

Jae-Joon Kim	POSTECH, South Korea

Special Sessions Chair

Gi-Joon Nam	IBM, USA

Local Arrangement Chairs

Ji-Hoon Kim	Chungnam National University, South Korea
Seokhyeong Kang	UNIST, South Korea

Publication Chairs

Yoonjin Kim	Sookmyung Women's University, South Korea
Jongeun Lee	UNIST, South Korea

Publicity Chairs

Tsung-Yi Ho	National Chiao Tung University, Taiwan
Nak Woong Eum	ETRI, South Korea
Hiroshi Nakamura	University of Tokyo, Japan
Jose L. Ayala	Complutense University of Madrid, Spain

Registration Chair

Jaeyong Chung   Incheon National University, South Korea

Finance Chair

Youngmin Yi   University of Seoul, South Korea

PhD Forum Chairs

Srinivas Katkoori   USF, USA
Jason Xue   City University of Hong Kong, Hong Kong, China

VLSI-SoC Steering Committee

Manfred Glesner   TU Darmstadt, Germany
Matthew Guthaus   UC Santa Cruz, USA
Salvador Mir   TIMA, France
Ricardo Reis   UFRGS, Brazil
Michel Robert   University of Montpellier, France
Luis Miguel Silveira   INESC ID/IST - University of Lisbon, Portugal
Chi-Ying Tsui   HKUST, Hong Kong, China
Fatih Ugurdag   Ozyegin University, Turkey

Technical Program Committee

Analog and Mixed-Signal IC Design

Chairs

Jaeha Kim   Seoul National University, South Korea
Tai-Cheng Lee   National Taiwan University, Taiwan

Members

Ke-Horng Chen   National Chiao-Tung University, Taiwan
Kenichi Okada   Tokyo Institute of Technology, Japan
Sai-Weng Sin   University of Macau, China
Michiel Steyaert   KU Leuven, Belgium
Jose M. de La Rosa   Instituto de Microelectrónica de Sevilla, IMSE-CNM
                             (CSIC), Spain
Jaehyouk Choi   Ulsan National Institute of Science and Technology,
                             South Korea

System Architectures NoC, 3D, Multi-core, and Reconfigurable

Chairs

Yuan Xie   UC Santa Barbara, USA
Nam Sung Kim   University of Wisconsin, USA

Members

Jishen Zhao   University of California, Santa Cruz, USA
Jiang Xu   Hong Kong University of Science and Technology,
                            Hong Kong, USA
Myoung Jung   UT Dallas, USA
Ulya Karpuzcu   University of Minnesota, USA
Radu Teodorescu   Ohio State University, USA
Leandro Indrusiak   University of York, USA
Ian O'Connor   Lyon Institute of Nanotechnology, France
Michael Huebner   Ruhr-University Bochum, Germany

CAD Synthesis and Analysis

Chairs

Minsik Cho   IBM, USA
Masahiro Fujita   University of Tokyo, Japan

Members

Bei Yu   UT Austin, USA
Duo Ding   Oracle Microelectronics, USA
Myung-Chul Kim   IBM Corporation, USA
Takashi Kambe   Kinki University, Japan
Tiziano Villa   Università di Verona, Italy
Ricardo Reis   Universidade Federal do Rio Grande do Sul, Brazil
Zhiru Zhang   Cornell University, USA

Circuits and Systems for Signal Processing and Communications

Chairs

Oscar Gustafsson   Linköping University, Sweden
Per Larsson-Edefors   Chalmers University, Sweden

Members

Hyeon-Min Bae   KAIST, South Korea
Liam Marnane   University College Cork, Ireland
Tobias Noll   RWTH Aachen University, Germany
Jongsun Park   Korea University, South Korea

Christoph Studer   Cornell University, USA
Dajiang Zhou   Waseda University, Japan
Fatih Ugurdag   Ozyegin University, Turkey
Luc Claesen   Universiteit Hasselt, Belgium

Embedded System Architecture, Design, and Software

Chairs

Vijaykrishnan Narayanan   Penn State University, USA
Jason Xue   City University of Hong Kong, Hong Kong, China

Members

Ingchao Lin   National Cheng Kung University, Taiwan
Wang Yu   Tsinghua University, China
Zili Shao   Hong Kong Polytechnic University, Hong Kong, China
Lar Bauer   Karlsruhe Institute of Technology, Germany
Koji Inoue   Kyushu University, Japan
Sri Parameswaran   University of New South Wales, Australia
Akash Kumar   National University of Singapore, Singapore

Low-Power and Thermal-Aware Design

Chairs

Massimo Poncino   Politecnico di Torino, Italy
Tadahiro Kuroda   Keio University, Japan

Members

Jose L. Ayala   Complutense University of Madrid, Spain
Aida Todri-Sanial   French National Center for Scientific Research, France
Mirko Loghi   Università di Udine, Italy
Donghwa Shin   Yeungnam University, South Korea
Chia-Lin Yang   National Taiwan University, Taiwan
Masaaki Kondo   The University of Electro-Communications, Japan

Memory Technology, Circuit, and System

Chairs

Yiran Chen   University of Pittsburg, USA
Rahul Rao   IBM, India

Members

Minki Cho   Intel, USA
Swaroop Ghosh   Intel, USA
Jingtong Hu   Oklahoma State University, USA

Kwanyeob Chae	Samsung Electronics, South Korea
Nitin Chandrachoodan	IIT Madras, India
Chengmo Yang	University of Delaware, USA
Lionel Torres	LIRMM, France

Prototyping, Verification, Modeling, and Simulation

Chairs

Graziano Pravadelli	University of Verona, Italy
Swarup Bhunia	Case Western Reserve University, USA

Members

Daniel Grosse	University of Bremen, Germany
Pierre-Emmanuel Gaillardon	Ecole Polytechnique Fédérale de Lausanne (EFPL), Switzerland
Anupam Chattopadhyay	Nanyang Technological University, Singapore
Prabhat Mishra	University of Florida, USA
Sandip Ray	Intel, USA
Laurence Pierre	TIMA, France
Florian Letombe	Synopsys, France
Adam Pawlak	Silesian University of Technology, Poland

Design for Variability, Reliability, and Test

Chairs

Chris Kim	University of Minnesota, USA
Jing-Jia Liu	National Tsing-Hua University, Taiwan

Members

Matteo Sonza Reorda	Politecnico di Torino, Italy
Swaroop Ghosh	Intel, USA
Victor Champac	INAOE, Mexico
Tony Kim	Nanyang Technological University, Singapore
Xiaofei Wang	University of Minnesota, USA
Satoshi Ohtake	Oita University, Japan

Security

Chairs

Ozgur Sinanoglu	New York University Abu Dhabi, UAE
Srinivas Katkoori	University of South Florida, USA

Members

Debdeep Mukhopadhyay	IIT Kharagpur, India
Mohammad Tehranipoor	University of Connecticut, USA
Paolo Maistri	TIMA Laboratory, France
Joseph Zambreno	Iowa State University, USA
Siddharth Garg	New York University, USA
Yier Jin	The University of Central Florida, USA

Contents

On the Use of System-on-Chip Technology in Next-Generation Instruments Avionics for Space Exploration

Xabier Iturbe[1]([⊠]), Didier Keymeulen[2], Patrick Yiu[3], Daniel Berisford[2], Robert Carlson[2], Kevin Hand[2], and Emre Ozer[1]

[1] ARM Research, Cambridge, UK
{xabier.iturbe,emre.ozer}@arm.com
[2] NASA Jet Propulsion Laboratory, Pasadena, CA, USA
didier.keymeulen@jpl.nasa.gov
[3] Massachusetts Institute of Technology, Cambridge, MA, USA
pyiu@mit.edu

Abstract. System-on-Chip (SoC) technology enables integrating all the functionality required to control and process science data delivered by space instruments in a single silicon chip (e.g., microprocessor + programmable logic). This chapter discusses the implications of using this technology in deep-space exploration avionics, namely in the next generation of NASA science instruments that will be used to explore our Solar system. We present here our experience at the NASA Jet Propulsion Laboratory (JPL) using Xilinx Zynq SoC devices to implement the data processing of a Fourier transform spectrometer, namely the Compositional InfraRed Imaging Spectrometer (CIRIS). Besides, we also discuss the different fault-tolerance techniques that have been implemented in the CIRIS controller SoC to deal with harsh radiation conditions prevailing in deep-space environments.

Keywords: Fault-tolerance · Avionics · System-on-chip integration · ARM processor · Signal processing

1 Introduction

Hybrid System-on-Chip (SoC) devices that embed the most energy efficient processor (ARM cores [1]) and the latest and most powerful FPGA architecture (Xilinx 7-series [2]) into a single chip (Xilinx Zynq [3]) promise new opportunities due to the performance, power consumption, weight and volume benefits they bring. This is especially relevant for building more capable space avionics.

Xabier Iturbe was also affiliated with the NASA Jet Propulsion Laboratory, California Institute of Technology, when conducting this research.
Patrick Yiu was affiliated with the California Institute of Technology when conducting this research.

Currently most of these systems combine programmable logic and processors as separate components distributed along one or several PCB board(s), which results in power consumption overheads and larger volume to be put into space [4,5]. Besides, currently existing space-grade processors (e.g., RAD750 [6]) are not suitable to be used in the next-generation spacecraft computing platforms because they do not provide sufficient performance and energy efficiency [7]. As a result, NASA and other space agencies have approached ARM and SoC technology, hoping to pave the way for future space exploration missions that are becoming ever more performance demanding.

Despite the fact that currently there are no space-qualified SoC parts, NASA is testing commercial Xilinx Zynq SoC devices in the International Space Station (ISS) as well as in precursor CubeSats operating in Low Earth Orbit (LEO), where the exposure to radiation is limited.

In view of a potential radiation-hardened SoC device that might be ready to fly in deep-space missions in the near to mid future, JPL and ARM have partnered together to develop a SoC platform to be used as a research vehicle for powering next-generation flight instruments intended to be used in NASA deep-space missions. Presently this platform, called APEX-SoC (APEX stands for Advanced Processor core for space EXploration), is being prototyped using a commercial Xilinx Zynq device. The APEX-SoC includes a generic and adaptable infrastructure that provides support for hardware and software based science processing. More specifically, the data acquisition and processing proper to each science instrument is to be implemented as a collection of "custom software and hardware applications" that are encapsulated by the APEX-SoC infrastructure and run on the Zynq's on-chip ARM processor and reside on the Zynq's FPGA fabric. Besides the infrastructure itself, the APEX-SoC includes a set of Radiation Hardened By Design (RHBD) features to protect the instrument-dependent modules implemented on the FPGA fabric from harsh space radiation. In connection with this, we are currently carrying out two research efforts to create a space-grade ARM processor that could potentially replace commercial ARM processors embedded in future radiation-tolerant SoC devices. First, we are conducting a thorough soft-error analysis of the ARM Cortex-R5 microprocessor, which is currently used in terrestrial safety-critical real-time applications, to identify the most vulnerable parts in the micro-architecture of this processor, analyze what level of protection is required for these vulnerable parts (e.g., detection only, correction only or hybrid), and then decide how to achieve this level of protection. Secondly, we are designing a Cortex-R5 based fail-operational Triple Core Lock-Step ARM processor (TCLS-ARM) with the capability to recover from errors within microseconds [8].

This chapter describes the first prototype of the APEX-SoC platform implemented on the Zynq SoC and presents an illustrative case-study drawn from the JPL Compositional Infrared Imaging Spectrometer (CIRIS) [11], which has been proposed to be used in icy moons, such as Jupiter's moon Europa [12]. The remainder of this chapter is as follows. Section 2 introduces the SoC technology and its use in space missions so far. Section 3 describes the APEX-SoC platform, and then Sect. 4 presents a case-study where the JPL CIRIS spectrometer data processing is implemented on this platform. Section 5 summarizes the implementation, performance and irradiation results that have been collected so far and, finally, Sect. 6 concludes the chapter and points out to future work.

2 System-on-Chip Technology and Its Use in Space Exploration Avionics

Despite miniaturized SoC technology is very convenient for space, where every gram of mass launched involves enormous costs, it is currently designed for and used by consumer terrestrial applications, where a single device with very low power consumption has found a niche in the network and telecommunication markets. Commercial SoCs have developed very advanced computation capabilities in consumer electronics that is continuously demanding more powerful devices and applications. One example of the sophistication degree achieved by commercial SoCs is the Xilinx Zynq-UltraScale+ MPSoC that is scheduled for release in early 2016 [13]. This will include an ARM Cortex-A53 high-performance 64-bit processor, an ARM Cortex-R5 real-time processor and a Xilinx UltraScale FPGA architecture.

Current SoC devices available in the market typically include at least one processor and an FPGA fabric. Since ARM cores are the standard processors used in all SoCs, the difference between them comes from the FPGA fabric they use. This fabric embeds routing resources, programmable logic, DSP and RAM blocks together with the memory cells to store their configuration.

Although the current use of SoCs is largely limited to terrestrial applications, space agencies consider this technology could be an alternative to overcome the current performance crisis seen in the space sector [7] as long as it develops an adequate degree of reliability to operate in harsh space environments. Indeed, when used in space, both ARM cores and FPGA fabric embedded in SoCs are vulnerable to radiation-induced soft-errors [9,10], which pose a greater reliability threat to SRAM-based FPGAs, such as those from Xilinx and Altera. In the latter FPGAs, the charged particles and outer radiation in general can alter the configuration information stored in SRAM-based memory cells, resulting in undesired logic functions implemented in the programmable logic and/or wrong inter-connections between the components. On the other hand Microsemi uses flash memory in its FPGA fabric, which is more resilient to radiation provoked soft-errors but allows for lower integration density, thus delivering more modest computation capabilities. Scrubbing is a classical method to protect the configuration memory in SRAM-based FPGAs. This technique consists in periodically checking the Error Correction Codes (ECCs) associated to the configuration information stored in the FPGA configuration memory and correct any errors that might have been occurred by rewriting the correct value, which is typically stored in an external rad-hard non-volatile flash memory. That said, Xilinx has released several generations of radiation-hardened FPGAs (e.g., Virtex-5QV [14]) and software tools for making designs fault-tolerant (e.g., Xilinx TMR Tool

[15]) that are used in a number of space systems. The Xilinx roadmap includes the development of a radiation-tolerant SoC technology as well as the necessary software tools for creating fault-tolerant designs on it.

Current space instrument payload systems typically include either a flash-based FPGA (e.g., Microsemi ProASIC3 [16]) or a rad-hard SRAM-based FPGA (e.g., Xilinx Virtex5-QV) for implementing data acquisition, synchronization and processing, and an antifuse-based FPGA (e.g., Microsemi RTAX [17]) for implementing data communications with spacecraft main computer, internal bus handling, housekeeping data collection and management of the configuration of the SRAM-based FPGA. In applications that are critical for spacecraft mission, such as Guidance Navigation and Control (GNC), the antifuse FPGA is replaced by a rad-hard processor such as a BAE Systems RAD750 [6] or a Cobham Gaisler Leon3 [18]. A couple of recent NASA instruments that use this classic architecture are the ChemCAM on the Mars Curiosity rover [4] and the Goddard Space Flight Center (GSFC) SpaceCube [5]. Hence, a SoC that includes these two components (processor + programmable logic) into a single chip is perfectly suited for space instrument payload systems.

Two are the reasons that have made us choose Xilinx Zynq SoC to proto-type our APEX-SoC platform. First, Xilinx is one of the vendors with the most advanced SoC technology roadmap, which also addressed radiation-hardened FPGAs. Second and most important, NASA has recently approached Xilinx technology in the scope of its CubeSat Launch initiative (CSLI), as described in the paragraph below. Xilinx Zynq SoCs integrate a dual-core ARM Cortex-A9 centric Processing System (PS) and a 28 nm Xilinx 7-Series (Artix-7 or Kintex-7) Programmable Logic (PL) fabric. The chip includes abundant on-chip AXI ports with low power rails to communicate the PS with the PL, which results in substantially less power consumption, considerably higher bandwidth and lower latency.

The Xilinx Zynq SoC is in the heart of the Computer Space Processor (CSP) designed by the National Science Foundation (NSF) Center for High-performance Reconfigurable Computing (CHREC) and licensed for fabrication to Space Micro Inc. [19–21]. The CSP uses a combination of commercial and rad-hard components, where commercial devices perform critical computations and are supervised by the rad-hard devices (e.g., reset and watchdog circuits). This Zynq-based processor will be part of future NASA missions such as the Space test Program-Houston-ISS-5 SpaceCube experiment [22] and the Compact Radiation bElt Explorer (CeREs) heliophysics CubeSat [23]. PlanetiQ Inc. will also integrate 3 CSPs on each of the 12 LEO weather satellites scheduled to be launched in 2017. In addition to these space missions, the CSP has been tested in neutron radiation and heavy-ion environment by Brigham Young University [24]. JPL, Xilinx and Swift LLC have also tested the Xilinx SoC part and other Xilinx 7-series FPGAs under heavy-ions radiation [25–27].

3   The APEX-SoC Platform and Infrastructure

The APEX-SoC platform is currently prototyped on a ZedBoard mini-ITX board, which is populated with a Xilinx Zynq 7Z100 SoC device. An FMC board containing an ADC is attached to the ZedBoard to deal with the analog electrical signals that are typically delivered by space science instruments. The APEX-SoC platform is also coupled with two external DDR memories to enable its use with instruments that generate large amounts of data: the PS-DDR is solely dedicated to the ARM processor in the Zynq, while the PL-DDR is used as scratchpad memory by the data processing modules implemented on the FPGA fabric and is also accessible by the ARM processor to retrieve the intermediate results computed by these. The last external component connected to the APEX-SoC is a SATA Solid State Device (SSD). This is used to temporarily store the (likely large amounts of) science results produced by the APEX-SoC until a downlink communication window with Earth is available, allowing for creating independent and stand-alone instruments avionics subsystems. The typical data-flow in the APEX-SoC is thus as follows: (1) the instrument data is acquired and processed by the FPGA logic, (2) the computed intermediate results by the FPGA logic are DMA-transferred to the DDR memory dedicated to the ARM processor for final processing, and (3) the final results are copied to the SSD prior to being downloaded to Earth.

The APEX-SoC provides support for integrating multiple identical data processing stages that can be used to process different science data in parallel to increase performance, or to detect computation errors by comparing their results when they process the same science data. This flexibility is needed when the requirements might change during the mission.

Figure 1 shows a block diagram of the APEX-SoC architecture. The following subsections describe the major aspects related to this architecture as well as the main fault-tolerance mechanisms that are implemented on it.

3.1   ARM-Centric Processing System

The ARM-centric PS includes all the peripherals that are typically required by flight science instruments, including: DMA support, GPIOs, Ethernet, SATA, interrupt controller and a memory-mapped register bank to exchange state and configuration data with the FPGA processing logic. As previously mentioned, process data are exchanged with the FPGA logic through the DMA-accessible PL-DDR memory. In order to speed-up the development of APEX-SoC-based instruments avionics, one of the ARM cores runs a standard Linux-based operating system, which provides Ethernet protocol to communicate with the spacecraft's main computer and a file system to ease the management of science results stored in the SSD. The second ARM core can be dedicated for software-based processing of instrument data. One scenario where software processing is convenient is when dealing with floating-point intensive algorithms, which can be easily computed using the NEON Floating Point Unit (FPU) [28] available in the ARM processor. A Real-Time Operating System (RTOS) can be deployed in

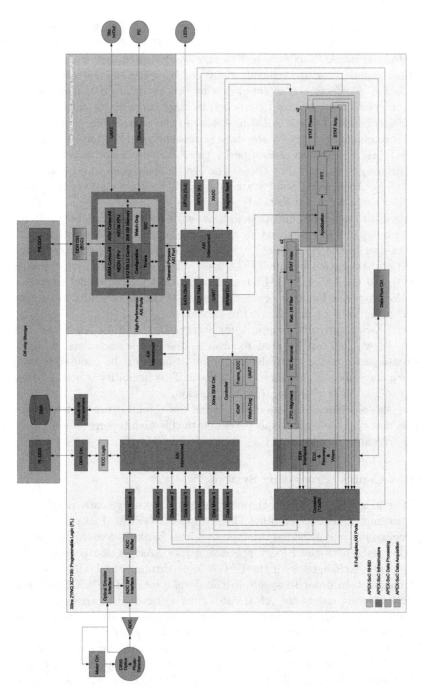

Fig. 1. The APEX-SoC platform

this core to use software multitasking to extend the hardware parallel processing carried out in the FPGA fabric while ensuring a sustainable use of CPU by all of the tasks [29].

3.2 Data-Flow Infrastructure

Each data processing module in the FPGA fabric is assigned a private data segment in the PL-DDR, with its size depending on the computing needs of that particular module. In order to exploit the full bandwidth delivered by the PL-DDR memory (6.4 GB/s) and to support the parallel/redundant execution of the hardware modules, the APEX-SoC implements eight 32-bit DDR access ports at 200 MHz using Xilinx-provided AXI-Stream Data Movers, which act as DMA controllers for the FPGA processing logic [30]. One of the DDR ports is dedicated to the instrument data acquisition logic (shown in blue color), another one is assigned to the ARM DMA, and the remaining six ports are connected to a crossbar that multiplexes them among the instrument data processing stages. The objective of this crossbar is thus to create as many communication channels as needed by the instrument-dependent modules using the physically available DDR ports. A data-flow controller drives the connections in the crossbar and schedules the PL-DDR accesses to maximize performance. For each data transfer, it specifies the memory address and size of the data segment to be read or written to the corresponding Data Mover. The data-flow controller is based on a tiny Xilinx 8-bit PicoBlaze processor [31], which consumes only 26 LUTs in the Zynq FPGA fabric, and implements a collection of reusable assembler routines that provide the required flexibility to deal with a wide range of instruments. Most of the HDL code used to describe the APEX-SoC infrastructure is also parameterizable and can be easily customized to the needs of any instrument.

3.3 Fault-Tolerance Features

The temperature on the Zynq die is continuously monitored using an on-chip sensor (see XADC in Fig. 1) [32] to identify and prevent overheat situations that could lead to the eventual destruction of the chip. Excessive noise situations in the power supply are also detected with this sensor. These may indicate that there is a problem with the voltage regulators, power lines in the PCB or even in the spacecraft power subsystem. Finally, the PL-DDR AXI Stream ports are continuously monitored to detect stuck-at situations and errors in memory data transfers. All storage resources in the APEX-SoC platform are protected with ECCs. The Xilinx ECC solution built in the silicon of the Zynq is used for the PS-DDR, whereas a custom ECC logic for the PL-DDR is implemented on the FPGA fabric. This ECC logic uses Hamming (32, 26) codes to protect the data words transferred through each of the PL-DDR ports and is pipelined to maximize performance. It allows for detecting and automatically correcting single bit flips (e.g., radiation-induced SEUs) in a PL-DDR data word and detecting, but not correcting, double bit errors. Note that the possibility that multiple bit errors are accumulated in the same data word is small, as the ECC logic corrects every single bit flip that might have occurred in the short period of time data remains stored in the PL-DDR memory between consecutive write accesses. Data words affected by uncorrectable double bit errors can be either replaced by zeros or with the interpolated value of the two neighboring samples. All the finite state machines in the APEX-SoC are implemented using "one-hot" encoding, in such a way that radiation-provoked upsets in state flip-flops result in the state machine flow being redirected to an "illegal" state that signals the ARM processor the error situation. The correctness of the configuration data stored in the Zynq configuration memory is periodically checked by a Xilinx Single Event Mitigation (SEM) controller [33]. Single-bit upsets are automatically fixed by the Xilinx SEM, and in the event of a double bit upset, the ARM processor carries out a full reconfiguration of the FPGA fabric.

3.4   Reliability Mode

As previously introduced, the APEX-SoC permits to increase system reliability by using multiple identical data processing stages in an N-out-of-M scheme. The number of M redundant stages that can be implemented is only limited by the amount of FPGA resources available on the fabric and the energy budget, however Dual Modular Redundancy (DMR) or Triple Modular Redundancy (TMR) are typically used. In all cases, three redundant copies of the same science data are kept in the PL-DDR memory and replicated majority voters are connected both at the input and output of the M redundant processing stages as shown in Fig. 2. The input voters do not consider corrupted data that cannot be recovered using ECCs. When any of the output voters detect that all of its input results are different, a computation error is assumed and the processing of that science dataset is repeated. Computation errors can occur when radiation affects data registers and/or FPGA configuration [10]. While upsets in the data registers cannot be detected by the Xilinx SEM controller, these are automatically corrected when reloading the data to process again. The Xilinx SEM is still needed to deal with the corrupted configuration bits, as described in Sect. 3.3. The data-flow controller coordinates the access by the voters to the redundant data in the PL-DDR in a ping-pong fashion, so that the voted results do not overwrite the source data, in case the computation needs to be repeated.

4   Case-Study: APEX-SoC-Based Controller of the JPL CIRIS Spectrometer

This section describes a proof-of-concept SoC implementation of a controller for the JPL CIRIS spectrometer using the APEX-SoC platform.

4.1   The JPL CIRIS Spectrometer

CIRIS is one of the new generation JPL instruments proposed to search for life indicators in icy moons, such as Europa [12]. It is based on the COTS instrument prototype described in [34], and it a small, rugged and lightweight Fourier

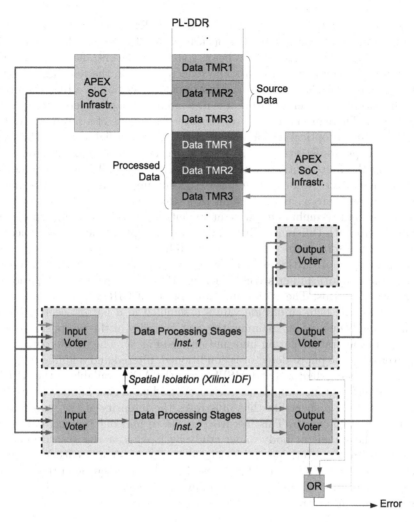

Fig. 2. DMR scheme implemented in the APEX-SoC-based CIRIS controller

Transform Spectrometer (FTS) with a high Signal-to-Noise Ratio (SNR) in the near-IR to thermal-IR region (2–12 µm) where the strongest and most diagnostic vibrational bands of the compounds of interest in Europa are found (e.g., 'CHNOPS' functional groups). CIRIS can work in cryogenic temperatures from 70–130 K with the use of passive cooling methods while onboard a spacecraft. More importantly, as opposed to related instruments such as grating spectrometers (e.g., Galileo NIMS [35]), CIRIS has intrinsic immunity from radiation-induced noise, enabling it to perform mid-IR solar reflectance and thermal emission spectroscopy with limited interference from the radiation environment in Europa.

The major structural novelty introduced by CIRIS is the constant-velocity rotating refractor it uses to vary the optical path difference of the two rays in which incoming light is divided by a beam splitter at the entrance of the instrument (red and green rays in Fig. 3). The reflected rays in the rotating refractor recombine after travelling through the instrument, resulting in a fringe interference light pattern (interferogram) that is measured with a photo-detector (purple ray in Fig. 3). There are up to four regions over the course of a revolution of the refractor where the optical interference between the input light rays can be measured with a photo-detector. These regions are located at approximately 16° arcs around the four positions where the refractor is parallel or perpendicular to the beam splitter. Note that the interferogram amplitude value is maximum in these four positions as all of the light rays travel the same distance along the spectrometer and recombine in phase at its output. This is why these positions are called Zero Path Difference (ZPD) positions. An optical incremental encoder mounted on the servomotor that drives the refractor's rotation is used on the ground prototype of CIRIS to identify these regions. As the CIRIS refractor performs 6.5 revolutions per second, each interferogram spans over a period of 13.6 ms every 24.8 ms. The optics and functioning of CIRIS result in an interferogram with the high-amplitude values assembled in a narrow central burst, and small-amplitude values spanning the vast majority of the tail positions and carrying the spectral resolution information (see Fig. 4). The interferogram signal delivered by the photo-detector is conditioned, filtered and amplified to ±5V range prior to being digitized at 1 MSPS using the ADC available in the APEX-SoC. The interferogram samples are then processed via a Fast Fourier Transform (FFT) to produce a spectrum that illustrates the intensity of the wavelengths present in the light beam. This in turn permits to find out the chemical composition of the sample or body under study by looking at the absorption lines in the spectrum. However, spectral leakage (e.g., "picket-fence" effect) and noise are also present in the spectrum due to the limited discretization of the interferograms through time limited digital sampling, and need to be properly handled by the instrument electronics to produce meaningful results [36].

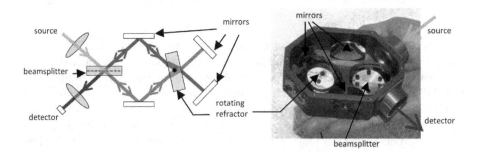

Fig. 3. CIRIS Spectrometer (Color figure online)

Although radiation has small impact on the spectral content of CIRIS data, FTS data processing allows increasing the SNR of the instrument even further [37]. As shown in Fig. 4, the shape of the CIRIS interferogram allows the data processing for detecting (and removing) most of the radiation hits that induce large current pulses (i.e., significantly greater than the nominal value) in the instrument's photo-detectors. In Fig. 4, note there are two radiation hits at -266 and $-500\,\mu s$.

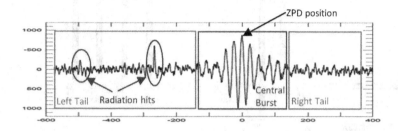

Fig. 4. CIRIS interferogram with radiation hits

At the moment there is a single photo-detector in the ground prototype of CIRIS, however the flight version of CIRIS will be equipped with an array of up to 25 photo-detectors to increase the instrument's spatial resolution and sensitivity in different IR bands. This will also increase the computation burden, as more interferograms will need to be processed within the same span of time (24.8 ms).

4.2   CIRIS Data Processing

The section describes the different processing stages that must be applied on the CIRIS interferogram data in order to produce meaningful spectroscopy results that can be interpreted by the scientists on Earth [36]. Figure 5 shows a block diagram of these stages as well as their interfaces with the APEX-SoC infrastructure. In this figure, note the two superposed main blocks that represent the dual data processing solution adopted in the CIRIS APEX-SoC to increase the performance and reliability.

The first stage prepares the digitized interferogram samples for subsequent processing by selecting 8,192 samples centered around the ZPD positions. This is done to deal with any temporal shift that might have occurred while sampling the interferogram.

The second stage removes the DC offset in the ZPD aligned interferogram by subtracting its average value, which is computed using a Cumulative Moving Average (CMA).

The third stage implements a radiation hit filter to detect and remove the outlier in the interferogram provoked by radiation striking the CIRIS photo-detector. The radiation pulses at the output of the CIRIS transconductance

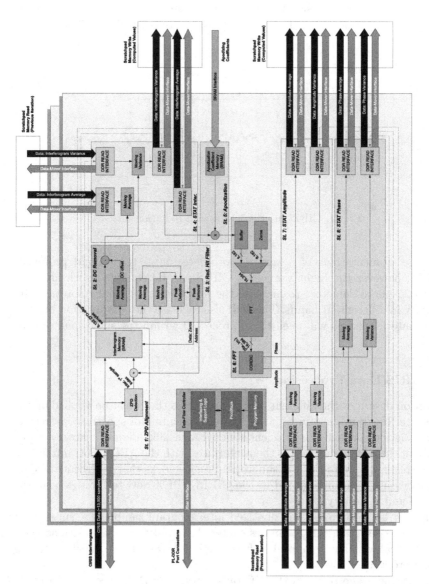

Fig. 5. CIRIS data processing block diagram

amplifier, with a bandwidth of 100 kHz, are about 10 µs full width at half maximum and easily recognized in the small-amplitude tail samples using statistics, namely the mean and variance (shown by triangles in Fig. 6) [37]. Radiation hit samples are then replaced by zeros without modifying significantly the spectral content of the interferogram. This property comes from the fact that interferogram points outside the central burst mainly carry redundant resolution information, and hence, removing a few points out of 8,192 lead to indistinguishable changes in the spectrum. In effect, each of the interferogram samples contributes only in about 0.1 % to the spectrum. Note here that the undetectable radiation hits that are at or below the un-irradiated noise level spread their energy over all wavelengths and therefore average to a constant DC offset in the spectrum, which is removed in the second stage. The mean and variance statistics are computed on the tail samples of the interferogram using the Knuth algorithm [38].

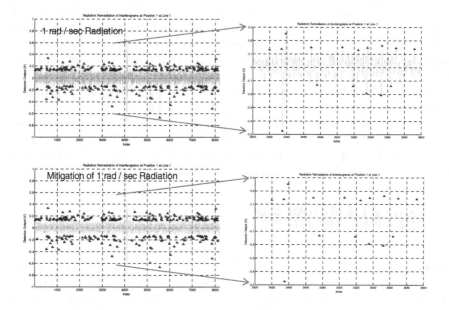

Fig. 6. CIRIS interferogram radiation effects and mitigation

The fourth stage (STAT Inter.) computes the variance and performs a CMA on successive interferograms detected around the same ZPD positions with the objective of estimating and increasing the SNR by removing the effect of high frequency and random noise. As in the third stage, the Knuth algorithm is used to calculate these statistics.

The fifth stage apodizes the averaged interferograms at the edges of the sampled regions to minimize the effects of spectral leakage.

The sixth stage computes the FFT on the interferogram. In light of increasing spectral resolution, this stage adds 4,096 zeros to each of the tails of the interferogram to obtain 8,192 additional interpolated spectrum points in-between the original nonzero-filled spectrum data, that is, 16,384 total spectrum points. This zero padding allows us to reduce the erroneous signal due to the "picket-fence" effect by up to 36 % [39].

The seventh stage relies on the Knuth algorithm to compute the variance and CMA on the spectrums resulting from the successive interferograms detected around the same ZPD positions.

The eighth and ninth stages are intended to correct the deviations provoked by CIRIS refractor's refractive index variations with wavelength and due to minor dissimilarities in the CIRIS optics between the four interferogram acquisition regions.

All of the processing stages described in this section are runtime configurable from Earth to adapt to potentially unexpected conditions when exploring distant planetary bodies. Some of the parameters that can be configured include: (1) the method to detect the position of the ZPD sample (e.g., most-negative, most-positive or most-magnitude) in the ZPD alignment core, (2) the number of trials to be averaged and the requirement to compute or not the variance in the STAT cores, (3) the apodizing function in the apodization core, and (4) the requirement for zero-filling (16,384 spectrum points) or not (8,192 spectrum points) in the FFT core.

It is important to note here that the interferograms detected in the photo-detector array that will be available in the flight version of CIRIS are independent of each other, and hence, the processing stages presented above are suitable for a parallel implementation on the APEX-SoC. Currently we simulate the photo-detector array by copying multiple times (to different PL-DDR memory segments) the same interferogram samples digitized by the single ADC in the system.

4.3   CIRIS Data Processing Integration into the APEX-SoC Infrastructure

All of the processing stages presented in Sect. 4.2 are implemented on the FPGA fabric, except stages 7 and 8, which run as software routines in the ARM processor because they involve floating-point operations. Two instances of the whole IRIS data processing are integrated into the APEX-SoC infrastructure, as shown in Fig. 5. As previously mentioned, these can be used to boost performance or to improve reliability (i.e., DMR scheme), depending on the mission's requirement at each time. For this specific scenario, a crossbar with 13 communication channels is created and the associated assembler routines in the data-flow controller are appropriately tailored for the data transfers required by CIRIS processing stages.

5   Results

This section summarizes the implementation, performance and irradiation results we have collected so far in the APEX-SoC-based CIRIS controller.

5.1   Implementation

The amount and type of resources consumed by the APEX-SoC infrastructure and the CIRIS data acquisition and processing modules when implemented on a Zynq 7Z100 device are detailed in Table 1. The most important implementation aspect to note here is the spatial isolation of the CIRIS modules within the FPGA fabric, which has been carried out following the Xilinx Isolation Design Flow (IDF) [40]. This permits to increase the availability of the system by preventing the situation where a single charged particle corrupts multiple processing stages in the FPGA fabric. As shown in Fig. 7, the crossbar and the data-flow controller are mapped in-between the two processing stages forming a fence. It is also important to note the small footprint of the voters compared to the processing stages, in the range of hundreds of LUTs and flip-flops, which minimizes the chances of being corrupted by radiation.

Table 1. Resources consumed in a Xilinx Zynq 7Z100 SoC

Component	LUTs	Flip-flops	DSP48s	BRAM36s
Data acquisition	347	267	N/A	N/A
RHBD features	6,194	3,818	N/A	14.5
Data processing	61,874	46,152	414	155
Infrastructure	61,773	49,631	N/A	395
Total	130,148 (47 %)	99,868 (18 %)	414 (20 %)	564.5 (75 %)

The power consumption reported by Xilinx Vivado design tool for the whole APEX-SoC-based CIRIS controller is approximately 5 W, which is about 3 W less than that of an equivalent board based controller.

The APEX-SoC uses up to 256 MB in the PL-DDR memory (approximately 25 % of the total DDR memory capacity) to process 25 interferograms simultaneously. The memory is arranged into several data segments across different categories, each containing a given type of data (e.g., raw interferogram, interferogram mean/variance, spectrum amplitude mean/variance or spectrum phase mean/variance) related to the information detected by a given photo-detector when the rotating refractor was in a given position. Besides, as explained in Sect. 3.4, each data is stored three times in the PL-DDR memory in different TMR data pools to increase reliability, and each TMR pool is itself replicated two times (A and B) to allow for data re-processing, if needed.

5.2   Performance

Table 2 shows the performance results measured when the FPGA processing logic is clocked at 200 MHz and both DDR memories and ARM Cortex-A9 processor run at 800 MHz. As shown in Fig. 8, the parallelism provided by the dual data

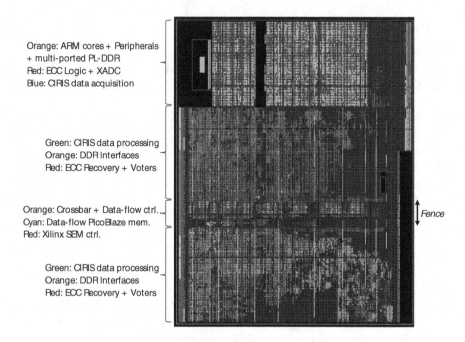

Orange: ARM cores + Peripherals
+ multi-ported PL-DDR
Red: ECC Logic + XADC
Blue: CIRIS data acquisition

Green: CIRIS data processing
Orange: DDR Interfaces
Red: ECC Recovery + Voters

Orange: Crossbar + Data-flow ctrl.
Cyan: Data-flow PicoBlaze mem.
Red: Xilinx SEM ctrl.

Green: CIRIS data processing
Orange: DDR Interfaces
Red: ECC Recovery + Voters

Fence

Fig. 7. APEX-SoC-based CIRIS controller floor-planning

processing channels in the APEX-SoC and the efficient PL-DDR memory access
schedule allow for processing two interferograms every 460 µs, with a latency
of 867 µs. Up to 4.6 GB/s of the total PL-DDR bandwidth (approx. 85 %) are
allocated to processing and the long latency introduced by the radiation hit
filter and the FFT computation is hidden by overlapping parallel processing
and PL-DDR data transfers. As a result, the APEX-SoC almost quadruples
the processing requirements of the flight CIRIS spectrometer as it is able to
process about a hundred interferograms within the time span the refractor in
the instrument takes to get between consecutive ZPD positions (24.8 ms). On the
other hand, when using the two redundant data processing channels to increase
reliability (i.e., DMR scheme), the APEX-SoC-based CIRIS controller is still able
to fulfill the processing requirement for the next-generation of CIRIS, requiring
up to 900 µs to process each interferogram.

5.3 Robustness Against Radiation

A radiation test was conducted at JPL using a 60Co γ-ray source (1 rad/sec)
directed toward the CIRIS photo-detector operating at 77 K to reduce detector
noise below the radiation hit pulses. The hit rate in this test was approximately
3,400 hits per second, exceeding what is expected in the Europa mission by at
least a factor of three. Figure 9 shows the obtained results, where the blue line
represents the measured values without radiation, the purple line represents the

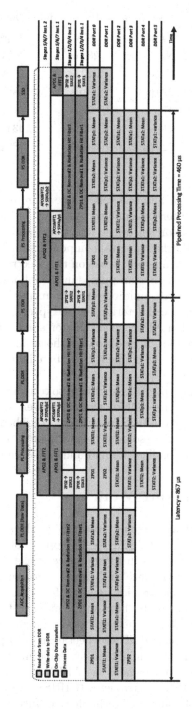

Fig. 8. Timing diagram and PL-DDR schedule (STATi=STAT Interferogram, STATa=STAT Amplitude STATp=STAT Phase)

Table 2. Performance results

CIRIS processing	Receive data	Process data	Deliver data
ZPD alignment	100 µs	N/A	41 µs
DC removal	N/A	281 µs	N/A
Radiation hit filter	N/A	320 µs	N/A
STAT interferogram	229 µs	~0 µs	188 µs/41 µs
Apodization	47 µs	~0 µs	41 µs
FFT	41 µs	165 µs	43 µs
STAT spectrum	229 µs	~0 µs	188 µs

values measured with radiation and the yellow line represents the values measured when the radiation hit filter was enabled. As shown in Fig. 9a, the radiation hit filter stage reduces the distortion of the line shape and spectrum provoked by radiation while keeping all other spectral components unaltered. Note in this figure that the high artificial "emission" peaks on the spectrum are coming from electrical noise generated by the vacuum pumps in the laboratory. These results are of utmost importance towards building an instrument that could cope with Europa-like radiation, which indeed deteriorated the spectroscopy data collected by NASA's previous generation NIMS spectrometer aboard the Galileo spacecraft more than a decade ago [35]. In addition, as shown in Fig. 9b, the adopted mitigation solution increases the instrument SNR by eliminating the noise due to radiation hit pulses.

We have not conducted any specific experiment to test the implemented fault-tolerance features yet, as these are well known and proven to be effective. Plans are to port the APEX-SoC-based CIRIS instrument described in this chapter to a radiation-hardened Xilinx SoC as soon as this technology is available and test the design in a simulated Europa-like thermal and radiation environment.

6   Conclusions and Future Work

This chapter has presented an ongoing research conducted by NASA's Jet Propulsion Laboratory (JPL) and ARM to develop a SoC platform (APEX-SoC) to power instruments avionics in future space exploration missions. This platform reduces significantly the size and power consumption of the instrument avionics as most of the electronics required for science processing of the instrument data are fitted in a single chip. At the moment this platform is prototyped using a commercial Xilinx Zynq SoC, where a number of fault-tolerance mechanisms have been implemented. The expectation is to port this design to a radiation-tolerant SoC part that might be available in the near future. The chapter has presented a case-study where the APEX-SoC prototype is used to process data delivered by a JPL spectrometer (CIRIS). Finally, the chapter has discussed the implications of using SoC technology in future space missions.

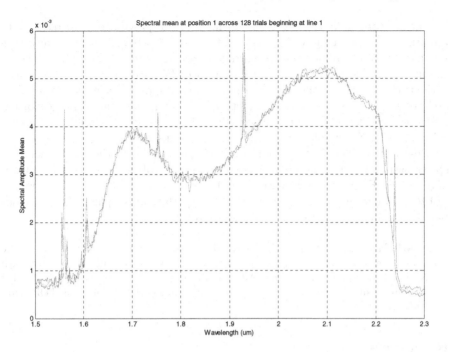

(a) Radiation mitigation in spectrum

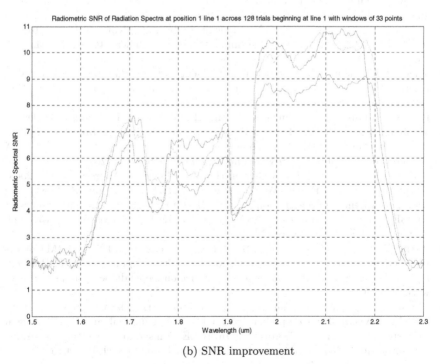

(b) SNR improvement

Fig. 9. Radiation mitigation in APEX-SoC-based CIRIS data processing

Future work at ARM will focus on making the processor more resilient to radiation. Namely, a thorough study of the ARM Cortex-R5 microarchitecture will be conducted to identify the parts that are more vulnerable and to choose the most suitable fault-tolerance techniques to be used in each of these parts without compromising the area and power consumption efficiency of the ARM architecture. At the processor architecture level, a fail-operational Triple Cortex-R5 Core Lock-Step (TCLS) processor will be developed [8]. JPL will look forward using the APEX-SoC platform with other space instruments.

Besides the research described in this chapter to design the next-generation space instruments avionics, JPL is also working in collaboration with the Goddard Space Flight Center (GSFC) and the Air Force Research Laboratory (AFRL) on designing a next-generation high-performance spaceflight processor based on a dual quad-core ARM Cortex-A53 [41].

Acknowledgment. The research described in this chapter was carried out at the Jet Propulsion Laboratory, California Institute of Technology, under a contract with the National Aeronautics and Space Administration (NASA). Xabier Iturbe is funded by the European Commission's FP7 Marie-Curie International Outgoing Fellowship Program with "Project No. 627579".

References

1. Blem, E., Menon, J., Sankaralingam, K.: A detailed analysis of contemporary ARM and x86 architectures. Technical report, University of Wisconsin - Madison (2013)
2. Mehta, N.: Xilinx 7 Series FPGAs: the logical advantage. Xilinx WP405 (2012)
3. Xilinx Inc.: Zynq-7000 All Programmable SoC: Technical Reference Manual, UG585 (2015)
4. Wiens, R.C., et al.: The ChemCam instrument suite on the Mars Science Laboratory (MSL) Rover: body unit and combined system tests. Space Sci. Rev. **170**, 167–227 (2012). Springer
5. Petrick, D., Gill, N., Hassouneh, M., Stone, R., Winternitz, L., Thomas, L., Davis, M., Sparacino, P., Flatley, T.: Adapting the SpaceCube v2.0 data processing system for mission-unique application requirements. In: Proceedings of the NASA/ESA Conference on Adaptive Hardware and Systems (AHS 2015) (2015)
6. BAE Systems Plc.: RAD750 Radiation-Hardened PowerPC Microprocessor (2008)
7. Doyle, R., Some, R., Powell, W., Mounce, G., Goforth, M., Horan, S., Lowry, M.: High performance spaceflight computing (HPSC) next-generation space processor (NGSP): a joint investment of NASA and AFRL. In: Proceedings of the Workshop on Spacecraft Flight Software (2013)
8. Iturbe, X., Venu, B., Ozer, E., Das, S.: A triple core lock-step (TCLS) ARM Cortex-R5 microprocessor for safety-critical and ultra-reliable applications. In: Proceedings of the IEEE/IFIP International Conference on Dependable Systems and Networks (2016)
9. Ebrahimi, M., Evans, A., Tahoori, M.B., Costenaro, E., Alexandrescu, D., Chandra, V., Seyyedi, R.: Comprehensive analysis of sequential and combinational soft errors in an embedded processor. In: IEEE Transactions on Computer-Aided Design of Integrated Circuits and Systems, vol. 34, no. 10, October 2015

10. Kastensmidt, F.L., Carro, L., Reis, R.: Fault-Tolerance Techniques for SRAM-Based FPGAs. Springer, Heidelberg (2006)
11. Berisford, D.F., Hand, K.P., Younse, P.J., Keymeulen, D., Carlson, R.W.: Thermal testing of the compositional infrared imaging spectrometer (CIRIS). In: Proceedings of the International Conference on Environmental Systems (2012)
12. Carlson, R.W., Hand, K.P., Berisford, D.F., Keymeulen, D.: The compositional infrared interferometric spectrometer (CIRIS) for assessing the habitability of Europa. In: Proceedings of the American Geophysical Union Fall Meeting (2013)
13. Xilinx Inc.: UltraScale Architecture and Product Overview, DS890 (2015)
14. Xilinx Inc.: Radiation-Hardened, Space Grade Virtex5-QV FPGA Data Sheet: DC and AC Switching Characteristics, DS692 (2015)
15. http://www.xilinx.com/ise/optional_prod/tmrtool.htm
16. Microsemi Inc.: Radiation-Tolerant ProASIC3 Low Power Spaceflight Flash FPGAs with Flash Freeze Technology (2012)
17. Microsemi Inc.: RTAX-S/SL and RTAX-DSP Radiation-Tolerant FPGAs (2015)
18. Cobham Gaisler: LEON3-FT SPARC V8 Processor Data Sheet and User's Manual (2013)
19. Rudolph, D., Wilson, C., Stewart, J., Gauvin, P., George, A., Lam, H., Crum, G., Wirthlin, M., Wilson, A., Stoddard, A.: CHREC space processor: a multifaceted hybrid architecture for space computing. In: Proceedings of the AIAA/USU Conference on Small Satellites (2014)
20. http://www.spacemicro.com/news/43.html
21. Mandl, D.: Intelligent payload module update. In: Proceedings of the HyspIRI Symposium (2015)
22. Wilson, C., Stewart, J., Gauvin, P., MacKinnon, J., Coole, J., Urriste, J., George, A., Crum, G., Timmons, E., Beck, J., Flatley, T., Wirthlin, M., Wilson, A., Stoddard, A.: CSP hybrid space computing for STP-H5/ISEM on ISS. In: Proceedings of the AIAA/USU Conference on Small Satellites (2014)
23. Kanekal, S., Jones, A., Randol, B., Patel, D., Summerlin, E., Gorman, E., Crum, G., Nolfo, G.D., Paschalidis, N., Heyward, S., Riall, S.: CeREs: a Compact Radiation bElt Explorer (2014)
24. Wirthlin, M.: Neutron radiation test results of the linux operating system executing within the CHREC space processor (CSP). In: Proceedings of the Military and Aerospace Programmable Logic Device International Conference (MAPLD 2015) (2015)
25. Amrbar, M., Irom, F., Guertin, S.M., Allen, G.: Heavy Ion single event effect measurements of Xilinx Zynq-7000 FPGA. In: Proceedings of the Radiation Effects Data Workshop (REDW 2015) (2015)
26. Swiftt, G.: Investigation of high current events in 28 nm 7-series FPGAs. In: Proceedings of the Military and Aerospace Programmable Logic Device International Conference (MAPLD 2015) (2015)
27. Koszek, W.A., Lesea, A., Steiner, G., White, D., Maillard, P.: Challenges in assessing single event upset impact on processor systems. In: Proceedings of the Workshop on Silicon Errors in Logic and System Effects (SELSE 2015) (2015)
28. ARM Ltd.: Cortex-A9 NEON Media Processing Engine: Technical Reference Manual (2011)
29. Xilinx Inc.: Zynq All Programmable SoC Linux-FreeRTOS AMP Guide, UG978 (2013)
30. Xilinx Inc.: AXI DataMover User Guide, PG022 (2014)
31. Chapman, K.: PicoBlaze for Spartan-6, Virtex-6, 7-Series, Zynq and UltraScale Devices (KCPSM6) (2014)

32. Xilinx Inc.: LogiCORE IP AXI XADC v1.00a Product Guide, PG019 (2012)
33. Xilinx Inc.: Soft Error Mitigation Controller, PG036 (2014)
34. Wadsworth, W., Dybwad, J.P.: Rugged high-speed rotary imaging fourier transform spectrometer for industrial use. In: Proceedings of the International Society for Optics and Photonics Conference on Environmental and Industrial Sensing (2002)
35. Carlson, R.W., Weissman, P., Smythe, W., Mahoney, J.: Near-infrared mapping spectrometer experiment on galileo. Space Sci. Rev. **60**, 457–502 (1992). Springer
36. Saptari, V.: Fourier Transform Spectroscopy Instrumentation Engineering, vol. 61. SPIE Press, Bellingham (2004)
37. Yiu, P., Iturbe, X., Keymeulen, D., Berisford, D., Hand, K.P., Carlson, R.W., Wadsworth, W., Levy, R.: Adaptive controller for a fourier transform spectrometer with space applications. In: Proceedings of the IEEE Aerospace Conference (2015)
38. Knuth, D.E.: The Art of Computer Programming: Seminumerical Algorithms, vol. 2. Addison-Wesley, Boston (1998)
39. Gronholz, J., Herres, W.: Understanding FT-IT data processing. Instrum. Comput. **3**(10), 1–23 (1985)
40. Hallett, E.: Isolation Design Flow for Xilinx 7 Series FPGAs or Zynq-7000 AP SoCs (ISE Tools), Xilinx XAp1086 (2015)
41. Statement of Work (SOW) for the Development of the High Performance Space Computing (HPSC) Processor. http://prod.nais.nasa.gov/eps/eps_data/167836-DRAFT-001-001.pdf

Fault Collapsing in Digital Circuits Using Fast Fault Dominance and Equivalence Analysis with SSBDDs

Raimund Ubar, Lembit Jürimägi[⊠], Elmet Orasson, and Jaan Raik

Department of Computer Engineering,
TTU, Ehitajate tee 5, 19086 Tallinn, Estonia
{raiub,elmet,jaan}@pld.ttu.ee,
lembit.jyrimagi@gmail.com

Abstract. The paper presents a new method and an algorithm for structural fault collapsing to reduce the search space for test generation, to speed up fault simulation and to make the fault diagnosis easier in digital circuits. The proposed method is based on hierarchical topology analysis of the circuit description at two levels. First, the gate-level circuit will be converted into a macro-level network of Fan-out Free Regions (FFR) each of them represented as a special type of structural BDD. This conversion procedure represents as a side-effect the first step of fault collapsing, resulting in a compressed Structurally Synthesized BDD (SSBDD) model explicitly representing the collapsed set of representative fault sites. The paper presents an algorithm which implements a complementary step of further fault collapsing. This algorithm is carried out at the macro-level FFR-network by topological reasoning of equivalence and dominance relations between the nodes of the SSBDDs. The algorithm has linear complexity and is implemented as a continuous scalable fault eliminating procedure. We introduce higher and lower bounds for fault collapsing and provide statistics of distribution of fault collapsing results over a broad set of benchmark circuits. Experimental research has demonstrated considerably better results of structural fault collapsing in comparison with state-of-the-art.

Keywords: Combinational circuits · Fault collapsing · Fault equivalence and dominance · Binary decision diagrams · Lower and higher bounds

1 Introduction

Fault collapsing is a procedure which is applied to reduce the number of faults of a given circuit to be targeted for testing purposes. Using a reduced set of only representative faults instead of a full set of faults has the goal to minimize the efforts in many test related tasks like test pattern generation, fault simulation for test quality evaluation, fault diagnosis, circuit testability evaluation etc.

The methods of fault collapsing are classified as structural and functional. Structural fault collapsing uses only the topology of the circuit whereas functional fault collapsing uses the circuit functional properties inherent in the circuit.

© IFIP International Federation for Information Processing 2016
Published by Springer International Publishing AG 2016. All Rights Reserved
Y. Shin et al. (Eds.): VLSI-SoC 2015, IFIP AICT 483, pp. 23–45, 2016.
DOI: 10.1007/978-3-319-46097-0_2

There are two classical ways used for structural fault collapsing: *fault equivalence* based and *fault dominance* based collapsing [1]. A fault f_j is said to *dominate* a fault f_i if every test that detects f_i also detects f_j. If f_j dominates f_i, only f_i needs to be considered during test generation. When two faults dominate each other, they are called *equivalent*. If two faults are equivalent, only one of them needs to be considered during test gene-ration or fault diagnosis. *Structural fault collapsing* uses the topology of the circuit structure. For example, a *stuck-at* 0 fault (SAF $y/0$) at the output y of AND gate is equivalent to all of the SAF $x/0$ faults at its inputs x_i. In a similar way, SAF $y/1$ at the output of AND gate dominates all the input SAF $x/1$ faults. The classical structural approaches to fault collapsing are based on gate-level circuit processing. An approach based on *fault-folding* was introduced in [2] for structural collapsing faults, using the iterative analysis of gate fault equivalence and dominance relations. Since structural fault collapsing is very fast, it is employed in many Automated Test Pattern Generators (ATPG) [3, 4].

Functional fault collapsing uses the circuit's functional information to establish equivalence and dominance relations. Two faults are functionally equivalent if they produce identical faulty functions [5] or we can say, two faults are functionally equivalent if we cannot distinguish them at the Primary Outputs (PO) with any input test vector [6]. Functional fault collapsing is generally regarded as very difficult to compute because it deals with the whole function of the circuit under test. In [7] it has been shown that the algorithmic complexity for identifying functionally equivalent faults is similar to that of ATPG.

Approximate fault collapsing via simulation has been proposed in [8]. In [9], a metric called *level of similarity* has been introduced and is efficiently used to improve the level of approximation. The fault collapsing suffers from the danger that if a fault in the collapsed fault set remains undetected then all other faults equivalent or dominating this fault removed from the collapsed fault set remain undetected as well. In [10], a safety parameter s to restrict the use of the dominance relation is introduced, and a safe fault collapsing method with a level of safety s is proposed.

The potentials of hierarchical fault collapsing were discussed in [11]. It was shown that hierarchical approach to fault collapsing gives more possibilities to increase the efficiency compared to the non-hierarchical one. An algorithm based on *transitive closures* on the *dominance graphs* has been proposed [12, 13], which enables more efficient hierarchical fault collapsing. It is a graph theoretic, fault independent and polynomial technique for functional fault collapsing.

In [14], functional dominance has been used to collapse the fault sets. However, this technique requires quadratic number of ATPG runs to obtain the collapsed fault set. An improvement was proposed in [15], which has the linear complexity regarding the number of ATPG runs. Since ATPG itself is used for learning functional dominance relations, both these techniques are suitable for small circuits only, but they can be helpful when combined with hierarchical fault collapsing. In [7] two theorems were introduced based on unique requirements and D-Frontiers of faults to extract equiva-lence and dominance relations. Similar approach was used in [16] based on the dominator theory for identifying more functionally equivalent fault pairs. In [17] a generalized dominance approach requires similar or lower run-times than that of [7].

A collapsed fault set helps generating smaller test sets for achieving the desired fault coverage, and it contributes to fault diagnosis as well. Since fault diagnosis deals with fault pairs, a linear reduction of the number of faults would result in a quadratic reduction of the target pairs.

In [5, 15], a novel diagnostic fault equivalence and dominance technique was proposed. A new method for fault collapsing for diagnosis called dominance with sub-faults was proposed in [18]. The method allows reducing the diagnosis search space. A framework where equivalence and dominance relations are defined for fault pairs is introduced in [19]. A fault pair collapsing is described, where fault pairs are removed from consideration under diagnostic fault simulation and test generation, since they are guaranteed to be distinguished when other pairs are distinguished. A technique to speed-up diagnosis via dominance relations between sets of faults using function-based techniques was proposed in [20]. Due to the high memory and time complexity this approach is applicable for small circuits only. All the listed techniques are fault oriented approaches, i.e. they consider a fault-pair at a time and use ATPG for identification of equivalence or dominance relations. In [10], a dynamic fault collapsing procedure is presented for fault diagnosis, where the faults are collapsed during the diagnostic test pattern generation contrary to the traditional static approaches described above where the faults are collapsed before test generation.

One of the main limitations of the described methods is that there is no evidence that investing more effort in fault collapsing reduces the total test generation time [10]. The reason is that most of the methods are using ATPG itself as a tool for fault collapsing, or they are usable only for small circuits because of the high computing complexity.

In this paper we concentrate on the structural fault-independent fault collapsing based on the topology analysis of the circuit. We target the minimal necessary set of representative faults as objectives for both, test generation and fault simulation. To cope with the complexity problem in case of big circuits, we use a hierarchical approach to structural fault collapsing, which is based on the topology analysis of the circuit at two levels – gate- and macro-levels, where the Fan-out-Free Regions (FFR) are regarded as macros. The proposed method is characterized at both levels by linear complexity which allows achieving high speed in fault collapsing, and provides smaller collapsed representative fault sets compared to other known structural methods. Due to low complexity, the method is well scalable and is therefore usable for large circuits where the functional fault collapsing methods give up because of the complexity.

The approach we propose consists of two consecutive procedures. During the first procedure, fault collapsing is carried out at the gate level by superposition of Binary Decision Diagrams (BDD) [21] of logic gates with the main goal of constructing a higher macro-level model of the circuit in form of Structurally Synthesized BDDs (SSBDD) [22, 23] where to each FFR an SSBDD corresponds. The fault collapsing can be regarded here as a side-effect (byproduct) of the SSBDD model synthesis. The second procedure, complementary part of the approach, is carried out at the higher macro-level by topological analysis of SSBDDs. Both parts of the fault collapsing procedure have linear complexity. It has been shown that SSBDDs can be efficiently used for fault simulation, outperforming in the speed state-of-the-art fault simulators [24, 25]. In this paper we show the possibility of additional fault collapsing using SSBDDs, which in turn can lead to further speed-up of fault simulation.

The paper is organized as follows. In Sect. 2 we give an overview of SSBDDs and in Sect. 3 we describe the synthesis of SSBDDs as the first step of gate-level fault collapsing. Section 4 presents the main theoretical concepts for the analysis of equivalence and dominance relations between the faults in the higher level FFR-networks modeled with SSBDDs, and Sect. 5 describes the algorithm of fault collapsing with SSBDDs. In Sect. 6, lower and higher bounds for fault collapsing are given. Section 7 presents experimental data, and Sect. 8 concludes the paper.

2   Structurally Synthesized BDD

Binary Decision Diagrams (BDD) have become by today a state-of-the-art data structure in VLSI CAD for representation and manipulation of Boolean functions. BDDs were first introduced for logic simulation in [26], and for test generation in [27, 28]. In 1986, Bryant proposed a new data structure called *Reduced Ordered BDDs* (ROBDDs) [21]. He showed simplicity of the graph manipulation and proved the model canonicity that made BDDs one of the most popular representations of Boolean functions. This model, however, suffers from the memory explosion problem, which limits its usability for large designs. Moreover, it cannot be used as a model for representing structural information about the design like representation of faults directly in the model. In [22, 27, 29], *Structurally Synthesized BDDs* (SSBDDs) were proposed with the goal to represent the structural features of circuits. The most significant difference between the function-based BDDs [21] and SSBDDs [22] is the method how they are generated. While BDDs are generated on the functional basis by *Shannon's expansions*, which handle only the Boolean function of the logic circuit, the SSBDD models are generated by a *superposition procedure* that extracts both, functions and data about structural signal paths of the circuit. The linear complexity of the SSBDD model results from the fact that a digital circuit is represented as a system of SSBDDs, where for each FFR a separate SSBDD is generated.

SSBDDs are generated by iterative superposition of library BDDs for simple or complex gates, guided by the structure of the given circuit. To avoid the explosion of the complexity of the SSBDD model, and to keep its size as minimal as possible, the superposition of BDDs is stopped at fan-out stems of the circuit. Using this restriction, to each FFR in the circuit an SSBDD will be created where a signal path in the FFR corresponds to each node in an SSBDD.

Example 1. An example of a combinational circuit and its SSBDD is depicted in Fig. 1. The SSBDD represents an FFR of the circuit obtained after cutting all the input fan-out branches of the circuit. This FFR can be described by the following Boolean expression:

$$y = f(X) = (x_1 x_{21} \lor (x_{22} x_3 \lor x_4 (\overline{x_5} \lor \overline{x_{61}})) \overline{x_{71}}) x_{81} \lor x_{82} x_9 (x_{72} \lor \overline{x_{62}}) \overline{x_{10}}$$

The non-terminal (internal) nodes of the SSBDD are labeled by the input variables of the FFR. To differentiate the fan-out branch variables from the fan-out stem variable we introduce for each of them a second subscript. The node variables may be inverted.

When using SSBDDs for calculating the output signals at given test patterns, we have to traverse the graph starting from the root node up to a terminal node guided by the input pattern. Let us agree that we exit each node during simulation to the right if the node variable has value 1, and downwards if the value is 0. In this case we don't need to label the edges in the graph by the values of the node variables on Figures. Entering the terminal node #1 as the outcome of graph traversing will mean the result of simulation $y = 1$, and entering the terminal node #0 will mean $y = 0$.

Example 2. For the circuit in Fig. 1 with function $y = f(X)$ the output signal y for the given input pattern X^t in Fig. 2 will be $y = 1$. During simulation of this pattern on the SSBDD, the following nodes are traversed: x_1, x_{22}, x_3, $\neg x_{71}$, x_{81}, #1 (shown by bold lines).

SSBDD model has several features that make it attractive compared to other commonly used mathematical models, such as conventional BDDs or gate-level netlists [30, 31]. The worst-case complexity (time) of generating SSBDD model from a circuit's netlist is linear in respect to the number of gates, while it is exponential for

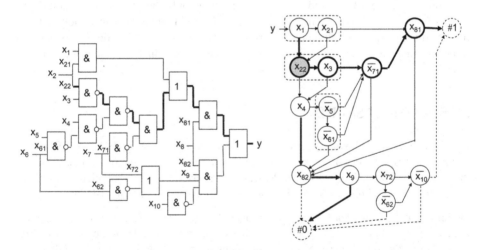

Fig. 1. Combinational circuit with a single output and its representation as an SSBDD

X	x_1	x_2	x_3	x_4	x_5	x_6	x_7	x_8	x_9	x_{10}	y
X^t	0	1	1	0	-	-	0	1	0	-	1

Tested: $x_{22} \equiv 0$, $x_3 \equiv 0$, $x_{71} \equiv 1$, $x_{81} \equiv 0$

Fig. 2. Test pattern for detecting selected faults SAF/0 or SAF/1 for the circuit in Fig. 1

common BDDs. The size of the SSBDD model is always linear in respect to the circuit size (BDDs can be of exponential size). Compared to the gate-level representation, SSBDDs help to reduce the complexity of the circuit by representing them as FFR (SSBDD) networks whereas the algorithms of processing the network components do not need dedicated treatment of the components described usually by design libraries. Moreover, instead of considering each gate separately, it deals with *macros* – FFRs represented by SSBDDs.

The most important feature of the SSBDD model is that it preserves structural information about the circuit while traditional BDDs do not. This is why differently from traditional BDDs, SSBDDs support structural test generation [22, 27] and fault simulation [24, 25, 29] for gate-level structural faults in terms of faulty signal paths with representing the faulty paths explicitly in the model. Each node in the SSBDD represents a signal path in the corresponding circuit, and the faults of the nodes represent the faults in signal paths.

For example, the SSBDD in Fig. 1 consists of 14 internal nodes where each of them represents a corresponding signal path of the total 14 paths in the circuit in Fig. 1 (the correspondence is shown by the variables x where x_i denote input signals, and x_{ij} denote the signals at the fan-out branches). The one-to-one mapping between the nodes in SSBDD and the paths in the circuit is the result of the SSBDD synthesis from the netlist of the given circuit. The synthesis process is presented in Sect. 3.

Note, that the SSBDD model in Fig. 1 represents only the FFR of the circuit. The faults of the input fan-out stems x_2, x_6, x_7 and x_8 should be handled separately, either by introducing trivial single-node BDDs to represent the input fan-out stems, or by modeling the stem faults as multiple faults in the nodes which represent the fan-out branches.

3   Synthesis of SSBDDs

Consider first, the following graph theory related definitions of the BDDs (SSBDDs). We use the graph theory notations instead of traditional *ite* expressions [21] because all the test related procedures based on SSBDDs are based on the topological reasoning rather than on symbolic manipulations as is traditionally the case for BDDs.

Definition 1. A BDD that represents a Boolean function $y = f(X)$, $X = (x_1, x_2, ..., x_n)$, is a directed acyclic graph $G_y = (y, M, \Gamma, X)$ with a set of nodes M and mapping Γ from M to M. $\Gamma(m) \subset M$ denotes the set of *successor* nodes of $m \in M$, and $\Gamma^{-1}(m) \subset M$ denotes the set of *predecessor* nodes of m. M consists of two types of nodes: *internal* (non-terminal) M^N and *terminal* M^T. For terminal nodes m^T we have $\Gamma(m^T) = \varnothing$. There is a single node $m_0 \in M$ where $\Gamma^{-1}(m) = \varnothing$ called the *root* node. A terminal node $m^T \in \{m^{T,0}, m^{T,1}\}$ is labeled by a constant $x(m^T) \in \{0,1\}$ and is called *leaf*, while all the nodes $m \in M^N$ are labeled by Boolean variables $x(m) \in X$, and have exactly two successor nodes $\Gamma(m) = \{m^0, m^1\}$.

Definition 2. We say, the edge $l(m, m^e)$ between nodes m and $m^e \in \Gamma(m)$ is *activated* when the node variable $x(m)$ is assigned to one of the values $e \in \{0,1\}$. We say, a path l

(m_i, m_j) between the nodes m_i and m_j is activated if all the edges which form the path are activated.

Definition 3. We say that a BDD $G_y = (y,M,\Gamma,X)$ represents a Boolean function $y = f(X)$, iff for every possible vector $X^t \in \{0,1\}^n$, a path $l(m_0, m^T)$ is activated so that $y = f(X^t) = x(m^T)$.

The main idea of superposition of BDDs as the basis procedure of SSBDDs proposed first in [22, 27], is illustrated in Example 3.

Example 3. Let us have in Fig. 3 a network of two components y and x_3 in Fig. 3, connected by the wire x_3. The components implement the following functions:

$$y = x_1x_2 \vee (x_3 \vee x_4)x_5, x_3 = x_6x_7 \vee x_8x_9$$

The components y and x_3 are represented by the SSBDDs y and x_3, respectively. For simplicity, we have omitted in SSBDDs the terminal nodes, with introducing the agreement that leaving the graph to the right means entering the terminal node #1, and leaving the graph down means entering the terminal node #0. Superposition of the two graphs y and x_3 is equivalent of merging the two components y and x_3 into a single component y* which implements the function:

$$y* = x_1x_2 \vee (x_6x_7 \vee x_8x_9 \vee x_4)x_5$$

To carry out this operation we have to substitute the node x_3 in the graph y with the graph x_3. To do that, we:

(1) connect the incoming edges of the node x_3 in graph y with the root node x_6 of graph x_3;
(2) connect all the nodes in the graph x_3, which enter into #1, with the right-hand neighbor of x_3 in graph y and

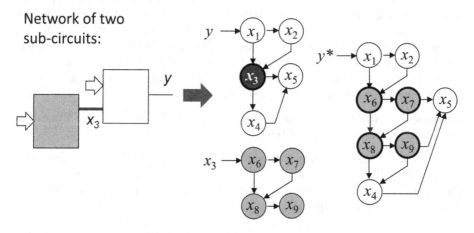

Fig. 3. Superposition of two SSBDDs

(3) connect all the nodes in the graph x_3, which enter into #0, with the down-hand neighbor of x_3 in graph y.

The new SSBDD y^* represents the function of the network with two merged components y and x_3.

Let us have, in general case, a gate-level circuit C where each gate is represented by an elementary BDD. The procedure of generating the SSBDD model $G(C)$ for C starts from the BDD of an output gate, and uses iteratively the superposition procedure where a node in a BDD is replaced by another BDD [22].

Consider two BDDs, G_y for the output gate $g_y \in C$ with output y, and G_x for the gate g_x connected to the input x of g_y. Let us call further, for simplicity, a node in a BDD labeled by a variable z as a "node z". By substitution of the node x in G_y with the BDD G_x we create from G_y a new SSBDD G_y' which represents now the extended network consisting of g_y and g_x. We call the new graph as SSBDD because the new nodes z in G_y' which belonged to G_x' represent the signal paths from the inputs z of the gate g_x via the connection line x between the two gates up to the output y of the gate g_y.

The procedure of the *superposition* of a node m labeled by x in BDD G_y with BDD G_x can be presented as follows.

Procedure 1. Superposition of BDDs

(1) The node m labeled by x is removed from G_y.
(2) All the edges in G_x connected to terminal nodes $m^{T,e}$ in G_x will be cut and then connected, respectively, to the successors m^e of the node m in G_y.
(3) All the incoming edges of m in G_y will be now incoming edges for the root node m_0 in G_x.

By applying *Procedure* 1 for two BDDs, we reduce the current model by one node and by one BDD. Suppose, the label variable x of a node m in a BDD G_y corresponds to the output of the gate g_x with k output branches. This means that the variable x is used as a label for k different nodes in the initial model as a set of BDDs. If we would proceed the superposition of graphs beyond the fan-out stem x, and would try to replace all the k nodes labeled by x with the BDD G_x, the complexity of the model would increase instead of reduction, i.e. the k nodes will be replaced by k BDDs G_x. Therefore, to keep the complexity of the final SSBDD model linear with the size of the circuit, and to reach the maximum compression of the initial model given as a set of elementary BDDs, we generate SSBDDs only for FFRs. Hence, at each fan-out stem we start a new superposition procedure for the next FFR.

Definition 4. A BDD which is constructed for a given FFR by Procedure 1 is called *structurally synthesized* BDD (SSBDD).

Corollary 1. It is easy to conclude from Procedure 1 that in the SSBDD G_y generated for the given FFR C_y with a function $y = f(x_1, x_2, ..., x_n)$, there are exactly n nodes with labels $x_1, x_2, ..., x_n$, and the node m with label x_i represents a unique signal path in C_y from the input x_i to the output y.

Corollary 2. Since all the SAF faults at the inputs of FFR according to the approach of fault folding [2] form the collapsed fault set of the FFR, and since all these faults are represented by the faults at the nodes of the corresponding SSBDD, then the creation of the SSBDD is equivalent to the fault collapsing procedure similar to fault folding.

Theorem 1. Let $G(C)$ be the SSBDD model generated for the combinational circuit C by Procedure 1. Then, any set of tests that checks all the SAF faults at the nodes of $G(C)$ checks all the SAF faults in C.

Proof. The proof follows from Corollaries 1 and 2, and from Theorem 5 in [2].

Unlike the traditional gate level approaches to test generation and fault simulation that use the collapsed fault list apart from the simulation model, the SSBDD based test generation and fault simulation are carried out on the macro-level (FFRs as macros) with direct representation of the faults in the model. Therefore there is no need for separate fault list to be used during test generation and fault simulation.

Example 4. The node x_{22} in the SSBDD represents the path from x_{22} to y in the circuit shown by bold lines in Fig. 1. On the other hand, the stuck-at faults SAF $y/0$ and SAF $y/1$ dominate the faults $x_{22}/0$ and $x_{22}/1$, respectively. The same dominance relation stands for all the faults along the bold path from x_{22} to y, regarding to the faults at x_{22}.

From this dominance relation, it results that all the faults along the signal path from x_{22} to y, except $x_{22}/0$ and $x_{22}/1$, can be collapsed. The two faults at x_{22} will form the representative fault subset for the full signal path from x_{22} to y. But, exactly these faults are represented in the SSBDD as the faults of the node x_{22}.

From above it follows that the SSBDD model can be regarded as the model where all the collapsed faults are removed and the fault sites are not visible either. This fault collapsing result is similar to that of fault folding method presented in [2].

The minimum size of SSBDDs, generated with Procedure 1 is always fixed and determined by the circuit structure. Let us denote N_{SSBDD} as the number of nodes in the SSBDD model, as the size characteristic of SSBDDs. Let $N_{Signals}$ be the number of lines, and N_G is the number of gates in the circuit represented by SSBDD. In [16] it has been shown that the number of nodes in SSBDDs can be calculated as

$$N_{SSBDD} = N_{Signals} - N_G$$

Since a digital circuit can be represented both by gate-level and by FFR-level SSBDDs, then in order to compare the gain in fault collapsing we get from translating the gate level SSBDD into FFR-level SSBDD, let us use the same units for measuring the size of SSBDDs in both cases – the number of SSBDD nodes in the model.

Denote s – as the number of inputs of the circuit, s_0 – as the number of inputs with no fan-outs, s_1 – as the number of internal lines with no fan-outs, s_k – as the number of nets in the circuit with k fan-outs ($k > 1$), n – as the number of outputs, and m – as the maximum number of fan-out branches over all fan-out stems in the circuit.

In [32] we have developed the following estimations for the sizes of SSBDDS for the gate-level N_{gate} and for FFR-level N_{SSBDD} cases:

$$N_{Gate} = s_0 + s_1 + \sum_{k=2}^{m} s_k(k+1)$$

$$N_{SSBDD} = s + \sum_{k=2}^{m} s_k(k+0)$$

Since in both cases for the stuck-at fault model, the number of nodes must be doubled to get the number of faults, then the ratio N_{gate} / N_{SSBDD} will characterize the gain in fault collapsing as the side-effect of FFR-level SSBDD synthesis from the initial gate-level SSBDD model, and the subtraction N_{gate} − N_{SSBDD} gives the exact number of collapsed faults thanks to the SSBDD synthesis.

Example 5. For the FFR of the circuit and its SSBDD in Fig. 1 we get N_{gate} = 30 and N_{SSBDD} = 18. The values of the arguments of the formulas for N_{gate} and N_{SSBDD} are depicted in Table 1. Hence, the gain in the SSBDD sizes, in this example, is 1.7, and the number of collapsed faults is 12. Note, the SSBDD with 14 nodes in Fig. 1 represents only the FFR part of the circuit. To get the full FFR-level SSBDD model, we have to include 4 single node SSBDDs for representing the 4 fan-out inputs in the circuit.

Table 1. Calculation of the number of nodes for 2 types of SSBDDs

	s_0	s_1	s_k	m	s	N
N_{gate}	6	12	4	2	–	N_{gate} = 6 + 12 + 4 * 3 = 30
N_{SSBDD}	–	–	4	2	10	N_{SSBDD} = 10 + 4*2 = 18

To summarize, the procedure of SSBDD synthesis can be regarded as the first part of fault collapsing for the given circuit. In the next section we will discuss the possibility of additional fault collapsing directly on the SSBDD model.

4   Fault Equivalence and Fault Dominance on the SSBDD Model

The second part of fault collapsing will consist of the processing of the SSBDD model with the goal to find additional set of faults which may be collapsed using the equivalence and dominance relationship on the SSBDD level. Since the nodes of SSBDDs represent signal paths on the gate-level circuit then each node related fault on the SSBDD to be collapsed is equivalent to all the related gate-level faults on the signal path represented by the node. Whereas the first part of fault collapsing was carried out by tracing the signal paths in the gate-level circuit level, then the second part concentrates on the path analysis at the higher FFR-level by tracing the paths on the SSBDDs.

Definition 5. Let us call a path $L(a, b)$ in the SSBDD between two nodes a and b, *activated* by a given input pattern X^t, if by traversing the graph under guidance of X^t, the node b will be reached from a.

In SSBDD-based test generation the targets are node related faults. As explained in [22], to test a node m in an SSBDD we have to activate three paths in it: (1) $L(m_0, m)$ from the root node m_0 to m, (2) $L(m^1, \#1)$ from the neighbor m^1 of m to the terminal node $\#1$, and (3) $L(m^0, \#0)$ from m^0 to $\#0$.

Example 6. To test the node x_{22} in the SSBDD in Fig. 1 we have to activate three paths in it: (1) $L(x_1, x_{22})$ from the root node x_1 to x_{22}, (2) $L(x_3, \#1)$ from x_3 to the terminal node $\#1$, and (3) $L(x_4, \#0)$ from x_4 to $\#0$. When we assign $x_{22} = 1$ then the activation of the listed paths produce a test pattern X^t which detects the fault SAF $x_{22} \equiv 0$. The pattern X^t which activates these paths (bold lines in Fig. 1) is depicted in Fig. 2.

Definition 6. Let us call the path which is activated from the root node up to the one of the terminal nodes, the *full activated path* in SSBDD. The full activated path which terminates in the node $\#1$ ($\#0$) is called 1-*path* (0-*path*). The nodes traversed along the 1-path (0-path) in direction to 1 (0), are called 1-*nodes* (0-*nodes*).

Example 7. The path $L(x_1, \#1) = (x_1, x_{22}, x_3, \neg x_7, x_{81}, \#1)$ in Fig. 1, activated by the pattern in Fig. 2, is 1-path, the node x_1 on this path is 0-node, and all other nodes are 1-nodes.

Property 1. If a test vector X^t activates in SSBDD a 0-path (1-path), then only 0-nodes (1-nodes) have to be considered as *candidate fault sites* [31].

The Property 1 can be taken into account to speed-up fault simulation. According to Property 1, the analysis of the 1-path in Fig. 2 shows us that all the nodes, except x_1, may be qualified as candidate fault sites. However, further analysis is needed to confirm which of the candidate nodes are in fact detectable by the pattern. Since the faults at all 1-nodes for X^t (in Fig. 2), will cause the direction change during graph traversing, then the faults at all 1-nodes are detectable by X^t.

Example 8. In the path $L(x_1, \#1)$ activated by the test in Fig. 2, according to Property 1, the nodes x_{22}, x_3, $\neg x_7$, and x_{81} are the candidates of fault sites. By additional simulation – by inverting the values of these variables, and by tracing the related paths $L(x_{22}, \#0)$, $L(x_3, \#0)$, $L(\neg x_7, \#0)$, $L(x_{81}, \#0)$ for each of these nodes, we can find that the test pattern in Fig. 2 detects the faults: $x_{22} \equiv 0$, $x_3 \equiv 0$, $\neg x_7 \equiv 0$, and $x_{81} \equiv 0$, respectively.

Theorem 2. The faults at two connected SSBDD nodes a and b are *equivalent* iff the following two conditions are satisfied: (1) the nodes have the same neighbor c, and (2) the node b has a single incoming edge from a.

Proof. The first condition refers to the fact that both nodes can be tested by the same test pattern which activates the paths $L(\text{Root},a)$, $L(a,\#e)$ where $e \in \{0,1\}$, and the path $L(c,\#(\neg e))$. The second condition refers to that this test pattern is the only one which can test both of the node faults $a/\neg e$ and $b/\neg e$.

Example 9. For example the faults $x_{22}/0$ and $x_3/0$ are equivalent, because the related nodes x_{22} and x_3 have the same neighbor node x_4, and a single entry edge into x_{22}, hence, one of these faults can be collapsed. In a similar way, using Theorem 2, it is easy to find in the SSBDD in Fig. 1 other equivalent faults: $x_1/0 \equiv x_{21}/0$, $x_5/0 \equiv x_{61}/0$ (or $\neg x_5/1 \equiv \neg x_{61}/1$, according to the notation in the SSBDD), $x_{8,2}/0 \equiv x_9/0$, and $x_{72}/1 \equiv x_{62}/0$. On the other hand, the faults $\neg x_{71}/0$ and $x_{81}/0$ are not equivalent. Despite of having the same neighbor x_{82}, the node $\neg x_{71}$ has three entry edges, and the single entry requirement of Theorem 2 is not satisfied.

Property 2. SSBDDs have always a single *Hamiltonian path* that visits all the nodes (except #0 and #1), and which determines a unique ranking of the nodes. The nodes a and b are in the relationship $a < b$ if the node a will be traversed before b along the Hamiltonian path [31].

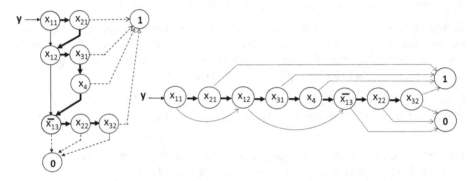

Fig. 4. Hamiltonian path in two presentations of the same SSBDDs

Figure 4 depicts an example of two possible presentations of the same SSBDD which represents the following Boolean expression:

$$y = x_{11}x_{21} \vee x_{12}(x_{31} \vee x_4) \vee \overline{x_{13}}x_{22}x_{32}$$

Theorem 3. The fault $b/0$ *dominates* $a/0$ (or $b/1$ *dominates* $a/1$), iff the following conditions are satisfied: (1) there exists a single 1-path (or a single 0-path) through the nodes for detecting both of these faults, (2) $a < b$, and (3) the node b has more than 1 incoming edges.

Proof. The first condition demands that these faults can be detected by a single test pattern (the condition of the equivalency). The second condition demands that there will be no other path for testing a and not testing b. The third condition is needed to give the possibility to test b and not to test a. From satisfying these conditions, it follows that any test for a must detect the related fault as well at b. Hence, the fault at a is dominated by b. If the third condition is not fulfilled, the related node faults at the nodes a and b are equivalent.

Example 10. In Fig. 1, the faults $\neg x_{71}/0$ and $x_{81}/0$ dominate $x_{22}/0$, and, according to Theorem 3, can be collapsed. Based on this result and taking into account Example 7, we can collapse 3 faults on the activated path $L(x_1,\#1)$: $x_3/0$, $\neg x_{71}/0$ and $x_{81}/0$.

Corollary 3. The fault $a/0$ dominates $b/0$ ($a/1$ dominates $b/1$) iff the following conditions are satisfied: (1) there exists a single 1-path (0-path) through the nodes for detecting both of these faults, (2) $a < b$, and (3) the node a can be tested by activating another path where b is not tested.

Proof. The proof results directly from Theorem 3 after transforming the SSBDD, so that the ranking of nodes a and b involved in the dominance relation will be swapped (see Fig. 5 and Example 10).

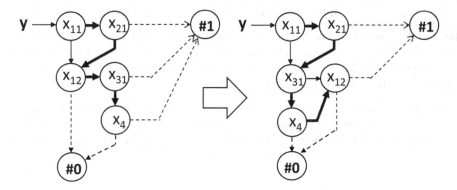

Fig. 5. Transformation of SSBDDs by swapping the nodes or subgraphs

Example 11. In Fig. 5 two different SSBDDs are shown which represent the same digital circuit, and correspond to the following two Boolean expressions:

$$y = x_{11}x_{21} \lor x_{12}(x_{31} \lor x_4) = x_{11}x_{21} \lor (x_{31} \lor x_4)x_{12}$$

The graphs represent the following rankings R1: $x_{12} < x_{31} < x_4$ and R2: $x_{31} < x_4 < x_{12}$, respectively, according to the Hamiltonian paths in the SSBDDs. In the SSBDD with node ranking R2, we determine that the node x_{12} dominates both, x_{31} and x_4, according to Theorem 3, and the same result we get for the SSBDD with node ranking R1, according to Corollary 3.

Using Theorems 2, 3 and Corollary 3 may directly lead to a simple algorithm of fault collapsing by systematic pairwise analysis of the equivalence and dominance relationships. However, in the worst case, such a pairwise analysis may lead to a quadratic complexity of SSBDD tracing.

On the other hand, taking into account the possibility of mapping sub-graphs in the SSBDD into the sub-circuits of the gate network, it would be possible to develop an algorithm of fault collapsing which will use only a single trace through the SSBDD with local analysis of proximate node pairs, and which would provide linear complexity of the algorithm.

5 Fault Equivalence and Fault Dominance Fast Reasoning on the SSBDD Model

From the definition of the SSBDDs [31] we can derive the following rules for recognition of gates and sub-circuits in the SSBDD model, which will help us to develop a fault collapsing algorithm on SSBDDs with linear complexity.

Definition 7. Let us call the consecutive nodes on the *Hamiltonian path* of SSBDD as a *group* if they all have the same neighbor node, and all these nodes except the first one have a single incoming edge.

Example 12. Consider a circuit and its SSBDD model in Fig. 1. The two consecutive nodes x_{22} and x_3, and the nodes $\neg x_5$ and $\neg x_{61}$ form two groups in the SSBDD in Fig. 3. No more groups exist in this graph. The nodes $\neg x_{61}$ and $\neg x_{71}$ don't form a group.

Rule 1. A group of two nodes connected by horizontal edges (vertical edges) represents AND (OR) gate, and due to the fault equivalence, a fault at one of the inputs can be collapsed. The Rule 1 results directly from the method of synthesis SSBDDs by superposition of BDDs of gates [22].

Example 13. The nodes x_{22} and x_3 in the SSBDD in Fig. 6 represent AND gate, and $\neg x_5$ with $\neg x_{61}$ represent OR gate. These gates can be recognized in the circuit. According to Rule 1, the faults $x_{22}/0$ (or $x_3/0$) and $\neg x_5/1$ (or $\neg x_{61}/1$) can be collapsed.

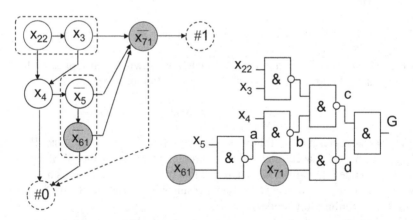

Fig. 6. Mapping SSBDD subgraphs into the circuit

Rule 2. If a node b in SSBDD has at least two or more incoming edges, it represents a path to a gate G where all the paths, represented by a subset of nodes $S(b) = \{a \mid a < b\}$, are joining. The fault of b dominates over the related faults of the nodes $a \in S(b)$, since the conditions of Theorem 3 are satisfied.

Example 14. The node $\neg x_{71}$ in SSBDD in Fig. 6 has three incoming edges. It represents a path to the gate G joining with the paths represented by all other nodes a,

$a < \neg x_{71}$, in this SSBDD. The nodes $\neg x_{61}$ and $\neg x_{71}$ don't form a group according to Definition 7 and don't represent AND.

The Rules 1 and 2 help to understand, how the fault equivalence and dominance relations in SSBDDs can be related to the similar equivalence and dominance relations in the gate-level circuit. If we have recognized a gate in SSBDD, the equivalence relations overlap for SSBDD and the circuit. The dominance relation in an SSBDD for a node with several incoming edges can be explained by transitive closure of dominance relations. For example, the dominance $\neg x_{71}/0 \rightarrow \neg x_{61}/0$ (or $x_{71}/1 \rightarrow x_{61}/1$ in the circuit) in the SSBDD in Fig. 6 can be explained by the following transitive closures in the circuit: $x_{71}/0 \equiv d/0 \equiv c/0$, and $c/0 \rightarrow b/1 \rightarrow a/0 \rightarrow x_{61}/1$, from which $x_{71}/1 \rightarrow x_{61}/1$ results ($x_{71}/1$ dominates $x_{61}/1$).

Algorithm 1 presents a procedure for fault collapsing in circuits which are represented by the SSBDD model. The algorithm is based on pairwise checking of Rule 1 (for equivalence) and Rule 2 (for dominance) by traversing along the Hamiltonian path in SSBDD. The algorithm has linear complexity.

Example 15. Consider the fault collapsing in the SSBDD in Fig. 1 according to Algorithm 1. The SSBDD represents the FFR region of the circuit in Fig. 1 where the fan-out inputs are not included. The initial number of the gate level SAF faults in the FFR in Fig. 1 is 52 (2 faults per each of 26 lines). By synthesizing the SSBDD for the FFR of the circuit according to Procedure 1 we reduce the number of representative faults from all 52 faults to 28 faults (2 faults per each of 14 nodes in the SSBDD). By using Algorithm 1, we further collapse 10 faults ($x_1/0$, $x_{22}/0$, $x_4/0$, $x_5/0$, $x_{71}/1$, $x_{81}/0$, $x_{82}/0$, $x_9/0$, $x_{72}/1$, $x_{10}/1$) which results in the total number of remaining 18 representative faults, i.e. 3 times reduction compared to the initial number of faults.

Table 2 shows which node faults in the SSBDD in Fig. 1 are collapsed on the basis of equivalence relation and which faults on the basis of dominance relation.

Table 2. Fault collapsing results for the SSBDD in Fig. 1

Node	Collapsed fault	Comments
x_1	SAF $x_1/0$	Equivalent with $x_{21}/0$
x_{21}	No collapse	
x_{22}	SAF $x_{22}/0$	Equivalent with $x_3/0$
x_3	No collapse	
x_4	SAF $x_4/0$	Dominates $x_5/1$, $x_{61}/1$
$\neg x_5$	SAF $x_5/0$	Equivalent with $x_{61}/0$
$\neg x_{61}$	No collapse	
$\neg x_{71}$	SAF $x_{71}/1$	Dominates $x_{22}/0$, $x_3/0$, $x_{22}/0$, $x_5/1$, $x_{61}/1$
x_{81}	SAF $x_{81}/0$	Dominates $x_{22}/0$, $x_3/0$, $x_{22}/0$, $x_5/1$, $x_{61}/1$, $x_1/0$, $x_{21}/0$, $x_{71}/1$
x_{82}	SAF $_{82}/0$	Equivalent with $x_9/0$
x_9	SAF $_9/0$	Dominates $x_{82}/0$
x_{72}	SAF $x_{72}/1$	Equivalent with $x_{62}/0$
$\neg x_{62}$	No collapse	
$\neg x_{10}$	SAF $x_{10}/1$	Dominates $x_{72}/0$, $x_{62}/1$, $x_9/0$, $x_{82}/0$

ALGORITHM 1: Fault collapsing on SSBDDs

Input: SSBDD model for a given circuit
Output: Set of collapsed faults C
Notations:
M — the number of all nodes
m — the number of the first node of the current node pair
n — the number of the second node, $n = m + 1$, $n* = n+1$
$d(m)$ — direction from node m to n
$n(d)$ — the neighbor of the node n in direction $d(m)$
$C(m)$ — the type of the collapsed fault $C(m) \in \{0,1\}$
$IN(m)$ / $OUT(m)$ — directions of incoming/outgoing edges
$FI(m)$ — flag to remember the multiple fan-in for m
$D(m)$ — flag to remember the direction of multiple fan-in

```
 1:  for all SSBDDs in the model
 2:    for all nodes m in the current SSBDD
 3:      if m < M then go to 6 end if
 4:      if m = M then C(m) = ¬IN(m) end if
 5:      go to 28
 6:      if m(¬d) = n(¬d) then                    (checking of Rule 1)
 7:        if FI(n) ≠ 1 then                      (checking of Rule 2)
 8:          if [(IN(m)>1) & (IN(m) = OUT(m))] or D(m) = 1 then
 9:            D(n) = 1
             end if
10:        if n = M then
11:          if IN(m)>1 or D(m) = 1 then C(n) = ¬d(m), go to 13 end if
12:          if OUT(m) ≠ OUT (n) then           (checking of Rule 1)
13:            C(m) = ¬d(m)
14:            if n(¬d) ≠ ∅ then FI(n(¬d)) = 1 end if
15:            m = m + 2, go to 3
             else
16:            if FI(n*) = 1 then go to 13 end if
17:            if n(¬d) = n*(¬d) then            (checking of Rule 1)
18:              C(m) = ¬d(m), m = m + 1, go to 3
19:            else C(n) = ¬d(m), go to 15 end if
             end if
           end if
20:      else go to 23 end if
21:    else if FI(m) = 1 then                    (checking of Rule 2)
22:        if FI(n) = 1 then                      (checking of Rule 2)
23:          C(m) = ¬IN(m)
24:          if m(¬d) ≠ ∅ then FI(m(¬d)) = 1 end if
25:          m = m + 1, go to 3
26:        else C(m) = ¬d(m), go to 24 end if
27:      else go to 26 end if
28:    end if
29:  end for
30: end for
```

6 Lower and Higher Bounds for Fault Collapsing

Denote by N the number of all nodes in the SSBDD model of a circuit and by C the number of collapsed faults. The number of all SAF faults is $2N$, the number of representative faults after fault collapsing will be $R = 2N - C$, and the effect from fault collapsing can be expressed by the ratio $R/2N$.

Theorem 4. The effect of fault collapsing in the SSBDD model of the given digital circuit will be always in the boundary $1/2 < R/2N \leq 5/6$. For a single FFR of any given digital circuit the effect of fault collapsing in the related single SSBDD will be always in the boundary $1/2 < R/2N \leq 3/4$.

Proof. Any tree-like circuit with N inputs can be represented by SSBDD with N nodes. Examples of such circuits, with fan-outs only in inputs, and the related SSBDDs for the FFRs are depicted in Figs. 7 and 8, respectively.

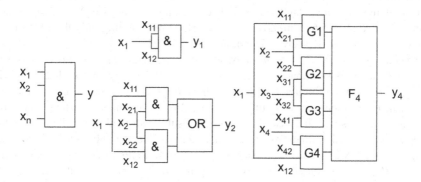

Fig. 7. Tree-like circuits with fan-out inputs of increasing complexity

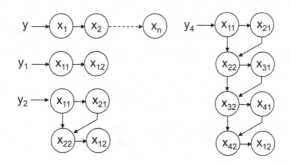

Fig. 8. SSBDD models for FFRs in the circuits in Fig. 4

In the simplest tree, a single gate with N inputs (gate y in Fig. 7 and SSBDD y in Fig. 8), we can collapse $N-1$ faults. Hence, $R = 2N - (N - 1) = N + 1$. Any partitioning of the set of inputs for more than one gate in this tree will reduce the total C by one fault per added gate and, hence, increase R. When increasing N, the lower bound for $R/2N$ is:

$$\lim_{n \to \infty} \frac{R+n}{2N+2n} = \lim_{n \to \infty} \frac{N+n+1}{2N+2n} = \lim_{n \to \infty} \frac{n}{2n} = \frac{1}{2}$$

On the other hand, consider formally (neglecting the redundancy) a single-input logic gate y_1 in Fig. 7. The SSBDD model of the gate has $N = 3$ nodes representing the fan-out stem with 2 branches. The SSBDD for only the FFR of this gate has 2 nodes. There are two equivalent faults at the gate inputs where one of them can be collapsed. Hence, the number of representative faults in this circuit will be $R = 2N - 1 = 5$, and $R/2N = 5/6$. Similarly, the SSBDD for the FFR has only $R = 2N - 1 = 3$, and $R/2N = 3/4$.

Consider now the tree-like circuit y_2 in Fig. 7 with two 2-input gates and two fan-out nodes. The SSBDD model of the circuit has $N = 6$ nodes. There are again two equivalent faults at the gate inputs where one of them can be collapsed. Hence, the number of representative faults after fault collapsing will be $R = 2N - 2 = 10$, and again we get $R/2N = 5/6$. Similarly, for the SSBDD representing only the FFR in this circuit, we get $N = 4$, $R = 2N - 2 = 6$, and $R/2N = 3/4$.

The circuit y_4 in Fig. 7 illustrates how we can generalize the series of two circuits y_1 and y_2 into a series of expanding circuits y_n, $n = 1,2,3,4...$, where each circuit will consist of an input sub-circuit IN_n as a chain of n 2-input gates, and a tree-like sub-circuit F_n. In each such a circuit, the ratio $R/2N = 5/6$ remains constant. In IN_n for each gate, only a single fault can be collapsed resulting in total in n collapsed faults.

It is easy to realize that any structural change inside the sub-circuit F_n will not change the ratio $R/2N = 5/6$. The reason is that all the faults in F_n will dominate the faults in IN_n. On the other hand, by adding $n = 1, 2,...$ non-fan-out inputs to the sub-circuit IN_n we get $R/2N^* = (R + n)/(2N + 2n)$, and by adding n fan-out inputs with 2 branches to IN_n we will get $R/2N^{**} = (R + 2n)/(2N + 6n)$. Each addition of a fan-out branch is equivalent to the case of adding a single input node where no faults can be collapsed.

From above it follows that for the case of digital circuits as networks of FFRs the higher bound (the worst case of remaining representative faults) for the ratio $R/2N$ will be:

$$R/2N^{**} < R/2N^* < R/2N \leq 5/6$$

Hence, the range between lower and higher bounds for the ratio $R/2N$ characterizing the number of remaining representative faults after using Algorithm 1 for fault collapsing will be:

$$1/2 < R/2N \leq 5/6$$

Extending the same analysis for a single sub-circuit as an FFR, we will have the range between lower and higher bounds for the ratio $R/2N$ as:

$$1/2 < R/2N \leq 3/4$$

Corollary 4. From Theorem 4, it directly follows that for the SSBDD model of the given digital circuit with N nodes in the model, the number of collapsed faults $C = 2N - R$ belong will always to the interval $N/3 \leq C < N$. Hence, $N/3$ will serve as the lower bound for the number of collapsed faults achievable in the SSBDD model.

Corollary 5. From Theorem 4, it directly follows that for the SSBDD with N nodes which represents any FFR in digital circuits, the number of collapsed faults $C = 2N - R$ will always belong to the interval $N/2 \leq C < N$. Hence, $N/2$ will serve as the lower bound for the number of collapsed faults achievable in any single SSBDD created for the given FFR.

Example 16. Consider again the digital circuit and the FFR of the circuit in Fig. 1. In Example 15 we found the number of collapsed faults $C = 10$, and the numbers of remaining representative faults for the case of FFR $R_{FFR} = 18$, and for the full circuits (with the SSBDD for FFR and additional 4 single-node SSBDDs for 4 inputs with fan-outs) R $_{circuit}$ = 26. For FFR we get $N/2 = 7 \leq \mathbf{C = 10} < N = 14$, whereas for the full model we have $N/3 = 6 \leq \mathbf{C = 10} < N = 18$.

The result shows that the interval between lower and upper bounds for a single FFR is smaller compared to the interval for whole circuit. More discussion in that topic follows in the experimental part of the paper.

7 Experimental Data

The fault collapsing experiments were carried out with Intel Core i5 3570 Quad Core 3.4 GHz, 8 GB RAM, using ISCAS'85, ISCAS'89 and ITC'99 benchmark circuits. The experimental results are presented in Tables 3 and 4.

In Table 3, the sizes of fault sets after fault collapsing for the proposed method (New) with previous structural [2, 14, 33] and functional [18] methods are compared. The new proposed method has better results in fault collapsing than the previous

Table 3. Comparison with other methods

Circuit	# Faults	Fault set size					CPU time, s	
		[2]	[14]	[33]	[18]	New	[18]	New
c1355	2710	1234	1210	1100	808	1210	46	0.003
c1908	3816	1568	1566	1286	753	1243	14	0.008
c2670	5340	2324	2317	2046	1853	1989	110	0.009
c3540	7080	2882	2786	2584	2092	2340	831	0.010
c5315	10630	4530	4492	4404	3443	3900	72	0.012
c6288	12576	5840	5824	4832	5824	5824	4	0.019
c7552	15104	6163	6132	5480	4707	5156	232	0.016

Table 4. Fault collapsing for ISCAS'89 and ITC'99 circuits

Circ	# Gates	R* [34, 35]	2N	R (New)	R/2N %	Gain R*/R	Time s
s13207	24882	9815	10456	7933	75.9	1.24	0.04
s15850	29682	11727	12150	9178	75.5	1.28	0.04
s35932	65248	39094	39094	29797	76.2	1.31	0.26
s38417	69662	31180	32320	25162	77.9	1.24	0.20
s38584	72346	36305	38358	28016	73.0	1.30	0.18
b15	47414	21072	23498	17439	74.2	1.21	0.04
b17	154220	68037	81330	60684	74.6	1.12	0.12
b18	463570	206736	277978	205866	74.1	1.00	0.42
b18_1	453088	202812	264244	196179	74.2	1.03	0.40
b19	1345442	533142	560704	415251	74.1	1.28	0.84
b19_1	1275720	507476	534184	396151	74.2	1.28	0.80
b21	79556	35994	48182	35169	73.0	1.02	0.08
b21_1	63732	29091	34510	25359	73.5	1.15	0.06
b22	113308	51277	70464	51511	73.1	1.00	0.11
b22_1	98006	44771	52172	38359	73.5	1.17	0.08
Aver	290392	121902	138643	102804	74.1	1.2	0.24

structural methods. The functional method [18] is very slow and not scalable due to high computational cost of calculating transitive closures on dominance graphs whereas the proposed method has a very high speed due to the linear complexity and is well scalable. As an example, the difference in time costs for c3540 and c6288 in case of [18] is 200 times whereas for the proposed method the difference is 2 times.

Due to different computing frameworks the speeds of the algorithms [18] and developed in this paper cannot be directly compared. On the other hand, the time cost needed for the first part of fault collapsing as a side effect of SSBDD synthesis was not included into the CPU time data in Table 3.

The experimental results for larger ISCAS'89 and ITC'99 circuits ($R*$ is the number of remaining faults after collapsing) are depicted in Table 4. The column $R*/R$ shows the gain (1.2 times in average) of the achieved fault collapse (in the column R(New)) compared to the results in [34, 35]. The last column shows that Algorithm 1 has linear complexity, is well scalable and can be efficiently used for large circuits. The linear complexity of the method is explained by the fact that the fault equivalence and dominance reasoning is reduced only to the local pairwise analysis of the neighbor nodes during traversing the Hamiltonian path of the SSBDD. The number of pairs to be analyzed, as it results from Algorithm 1, is in the interval $(N - 1, N/2 + 1)$ where the lower bound refers to the extreme case of the logic gate with N inputs, and the higher bound refers to the extreme case of the two-level AND-OR (OR-AND) circuits with 2 inputs for the 1st level AND (OR) gates, plus one additional input for the 2-nd level gate.

Note, that according to Theorem 4, the higher bound for R/2 N is 83 % and the lower bound is 50 %. The best result of fault collapsing – 73.0 % of remaining faults, the worst result – 77.9 % and the average of 74.1 % all fit well into the interval between the bounds. However the results are considerably closer to the higher bound of 83.3 % of remaining representative faults than the lower bound of 50 %.

In Fig. 9 we show statistical data collected from the fault collapse experiments with SSBDDs in 0.5 million tree-like sub-circuits (FFRs) in 111 different circuits of ISCAS'85, ISCAS'89 and ITC'99 families. Figure 6 presents a plot of different sub-circuit cases characterized by the number of nodes N in SSBDDs and the results of fault collapsing $R/2N$. Two extreme cases are highlighted: single-gate circuits (the best fault collapsing case) and the circuits with 2-input gates at the first level of tree-like circuits – the worst fault collapsing case where the higher bound for the remaining representative faults $R/2N = 3/4$ was reached, respectively the lower bound (the minimum number) of faults collapsing $C = N/2$.

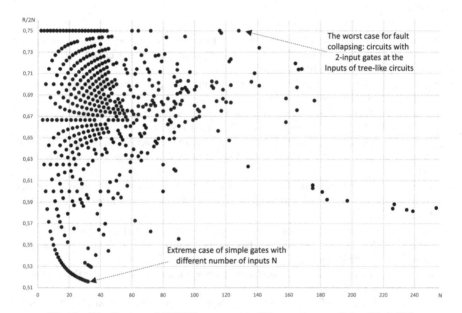

Fig. 9. Distribution of SSBDD cases with different characteristics (N, R/2N)

8 Conclusions

In this paper we proposed a new structural fault collapsing method and an algorithm with linear complexity. The method is based on using SSBDD model for representing gate-level circuits as higher FFR-level networks. The synthesis of SSBDDs presents the first step of fault collapsing in FFR-s, and the resulting collapsed fault set can be regarded as a side-effect of SSBDD synthesis. We have introduced the concepts of fault equivalence and dominance relations between the faults on the SSBDD model, and present an algorithm for systematic fault collapsing in SSBDDs as a process of creating of the representative fault set defined at the higher level communication network of FFRs.

We developed the lower and higher bounds of the SSBDD based fault collapsing, and showed that the number of collapsed faults C in the SSBDD model of an arbitrary digital circuits belongs to the interval $N/3 \leq C < N$ where N is the number of nodes in the SSBDD model.

Experiments showed that the proposed method is more efficient than the previous structural fault collapsing methods and due to high scalability makes it very promising for large circuits.

Acknowledgments. The work has been supported by EU FP7 STREP project BASTION, and HORIZON 2020 RIA project IMMORTAL.

References

1. Bushnell, G.M.L., Agrawal, V.D.: Essentials of Electronic Testing. Springer, Boston (2000)
2. To, K.: Fault folding for irredundant and redundant combinational circuits. IEEE Trans. Comput. **C-22**(11), 1008–1015 (1973)
3. Niermann, T., Patel, J.H.: HITEC: a test generation package for sequential circuits. In: EDAC, pp. 214–218, February 1991
4. Kelsey, T.P., Saluja, K.K., Lee, S.Y.: An efficient algorithm for sequential circuit test generation. IEEE Trans. Comput. **42**(11), 1361–1371 (1993)
5. Sandireddy, R.K.K.R., Agrawal, V.D.: Using hierarchy in design automation: the fault collapsing problem. In: Proceedings of the 11th VLSI Design and Test Symposium Kolkata, 8–11 August 2007
6. Veneris, A., Chang, R., Abadir, M.S., Seyedi, S.: Functional fault equivalence and diagnostic test generation in combinational logic circuits using conventional ATPG. JETTA **21**(5), 495–502 (2005)
7. Lioy, A.: Advanced fault collapsing. IEEE Des. Test Comput. **9**(1), 64–71 (1992)
8. Al-Assad, H., Lee, R.: Simulation based approximate global fault collapsing. In: Proceedings of International Conference on VLSI, pp. 72–77 (2002)
9. Pomeranz, I., Reddy, M.: Level of similarity: a metric for fault collapsing. In: Proceedings of DATE, pp. 56–61, February 2004
10. Pomeranz, I., Reddy, S.: Safe fault collapsing based on dominance relations. In: Proceedings of ETC, pp. 7–8 (2008)
11. Hahn, R., Krieger, R., Becker, B.: A hierarchical approach to fault collapsing. In: Proceedings of EDTC, pp. 171–176 (1994)
12. Prasad, A.V.S.S., Agrawal, V.D., Atre, M.V.: A new algorithm for global fault collapsing into equivalence and dominance sets. In: Proceedings of ITC, pp. 391–397, October 2002
13. Sethuram, R., Bushnell, M.L., Agrawal, V.D.: Fault nodes in implication graph for equivalence dominance collapsing, and identifying untestable and independent faults. In: Proceedings of VLSI Test Symposium, pp. 329–335 (2008)
14. Agrawal, V.D., Prasad, A.V.S.S., Atre, M.V.: Fault collapsing via functional dominance. In: International Test Conference, pp. 274–280 (2003)
15. Sandireddy, R.K.K.R., Agrawal, V.D.: Diagnostic and detection fault collapsing for multiple output circuits. In: Proceedings of DATE, pp. 1014–1019 (2005)
16. Amyeen, M.E., Fuchs, W.K., Pomeranz, I., Boppana, V.: Fault equivalent identification in combinational circuits using implication and evaluation techniques. IEEE Trans. CAD **22**(7), 922–936 (2003)

17. Vimjam, V.C., Hsiao, M.S.: Efficient fault collapsing via generalized dominance relations. In: Proceedings of VLSI Test Symposium, pp. 258–265 (2006)
18. Adapa, R., Tragoudas, S., Michael, M.K.: Sub-faults identification for collapsing in diagnosis. In: International Conference ISCAS, pp. 815–818 (2006)
19. Pomeranz, I., Reddy, S.: Equivalence and dominance relations between fault pairs and their use in fault pair collapsing for fault diagnosis. In: International Conference on VLSI Design, pp. 1–6 (2007)
20. Adapa, R., Tragoudas, S., Michael, M.K.: Accelerating diagnosis via dominance relations between sets of faults. In: Proceedings of the VLSI Test Symposium, pp. 219–224 (2007)
21. Bryant, R.: Graph-based algorithms for boolean function manipulation. IEEE Trans. Comput. **C-35**, 677–691 (1986)
22. Ubar, R.: Test synthesis with alternative graphs. IEEE Des. Test Comput. **13**, 48–59 (1996). Springer
23. Ubar, R., Raik, J., Vierhaus, H.-T.: Design and test technology for dependable systems-on-chip. In: IGI Global, p. 550 (2011)
24. Ubar, R., Devadze, S., Raik, J., Jutman, A.: Parallel X-fault simulation with critical path tracing technique. In: Proceedings of DATE (2010)
25. Gorev, M., Ubar, R., Devadze, S.: Fault simulation with parallel exact critical path tracing in multiple core environment. In: DATE (2015)
26. Lee, C.Y.: Representation of switching circuits by binary decision diagrams. Bell Syst. Tech. J. **38**(7), 985–999 (1959)
27. Ubar, R.: Test generation for digital circuits using alternative graphs. In: Proceedings of the Tallinn Technical University, Tallinn, vol. 409, pp. 75–81 (1976)
28. Akers, S.: Binary decision diagrams. IEEE Trans. Comput. **27**, 509–516 (1978)
29. Ubar, R.: Multi-valued simulation of digital circuits with structurally synthesized BDDs. In: Multiple Valued Logic, vol. 4. OPA, Gordon and Breach Publishers (1998)
30. Jutman, A., Raik, J., Ubar, R.: SSBDDs: Advantageous model and efficient algorithms for digital circuit modeling, simulation and test. In: 5th International Workshop on Boolean Problems, Freiberg, Germany, pp. 157–166, 19–20 September 2002
31. Ubar, R.: Overview about low-level and high-level decision diagrams for diagnostic modeling of digital systems. Facta Univ. (Nis) Ser.: Electron Energ. **24**(3), 303–324 (2011)
32. Mironov, D., Ubar, R.: Lower bounds of the size of shared structurally synthesized BDDs. In: IEEE 17th International Symposium on Design and Diagnostics of Electronic Circuits and Systems (DDECS), Warsaw, pp. 77–82, 23–25 April 2014
33. Ubar, R., Mironov, D., Raik, J., Jutman, A.: Structural fault collapsing by superposition of BDDs for test generation in digital circuits. In: IEEE ISQED, San Jose, CA, USA, pp. 250–257 (2010)
34. Brglez, F., et al.: Combinational profiles of sequential benchmark circuits. In: ISCAS 1989 (1989)
35. ITC 1999: http://www.cad.polito.it/downloads/tools/itc99.html

A Hardware Accelerator for Real Time Sliding Window Based Pedestrian Detection on High Resolution Images

Asim Khan$^{(\boxtimes)}$, Muhammad Umar Karim Khan, Muhammad Bilal, and Chong-Min Kyung

Department of Electrical Engineering,
Korea Advanced Institute of Science and Technology (KAIST),
Daejeon, South Korea
{asimkhan,umar,bilalm,kyung}@kaist.ac.kr

Abstract. Pedestrian detection has lately attracted considerable interest from researchers due to many practical applications. However, the low accuracy and high complexity of pedestrian detection has still not enabled its use in successful commercial applications. In this chapter, we present insights into the complexity-accuracy relationship of pedestrian detection. We consider the Histogram of Oriented Gradients (HOG) scheme with linear Support Vector Machine (LinSVM) as a benchmark. We describe parallel implementations of various blocks of the pedestrian detection system which are designed for full-HD (1920 × 1080) resolution. Features are improved by optimal selection of cell size and histogram bins which have been shown to significantly affect the accuracy and complexity of pedestrian detection. It is seen that with a careful choice of these parameters a frame rate of 39.2 fps is achieved with a negligible loss in accuracy which is 16.3x and 3.8x higher than state of the art GPU and FPGA implementations respectively. Moreover 97.14 % and 10.2 % reduction in energy consumption is observed to process one frame. Finally, features are further enhanced by removing petty gradients in histograms which result in loss of accuracy. This increases the frame rate to 42.7 fps (18x and 4.1x higher) and lowers the energy consumption by 97.34 % and 16.4 % while improving the accuracy by 2 % as compared to state of the art GPU and FPGA implementations respectively.

Keywords: FPGA · Low power · Object detection · Real-time

1 Introduction

Researchers in industry and academia have been striving for accurate and real-time pedestrian detection (PD) for more than a decade owing to many commercial and military applications. Industries such as surveillance, robotics, and entertainment will be greatly influenced by appropriate application of PD. Advanced driver assistance systems (ADAS) and unmanned ground vehicles (UGV) are merely a distant dream without automated pedestrian detection. The fact that more than 15 % of traffic accidents include pedestrians [1] shows the importance of real-time pedestrian detection for the modern society [2].

Y. Shin et al. (Eds.): VLSI-SoC 2015, IFIP AICT 483, pp. 46–66, 2016.
DOI: 10.1007/978-3-319-46097-0_3

Amid numerous applications, the search for an accurate yet fast PD algorithm is ongoing. Researchers have shown great interest over the past few years in extracting diverse features from an image and finding an appropriate classification method to perform robust PD [10–14]. However, the histogram of oriented gradients (HOG) approach has proven to be a groundbreaking effort, and has shown good accuracy in various illumination conditions and multiple textured objects. Inspired from SIFT [5], the authors in their seminal paper [4] present a set of features over a dense grid in a search window. For training and classification, they used the linear support vector machine (linSVM). Their work inspired many other researchers and is still used as a benchmark PD scheme.

Although HOG was presented many years back, it is surprising to see that very few efforts have been made for an optimal hardware implementation of HOG. In fact, most of the research has been targeting pedestrian detection on a high end CPU or GPU or combination of both [24–29]. Field Programmable Gate Arrays (FPGA) and Application Specific Integrated Circuits (ASIC) often provide better execution speed and energy efficiency as compared to GPUs due to deep pipelined architectures. Furthermore, in many embedded applications, such as surveillance, there are numerous constraints on hardware cost, speed, and power consumption. For such applications, it is more suitable to use task-specific (FPGA, ASIC) rather than general-purpose platforms. Moreover, to meet such constraints, certain parameters of the algorithm need to be tuned and an insight is required into how the change of parameters of PD affects not only the accuracy but also the hardware complexity.

Efforts have been made in the research community to either improve the accuracy of PD or reduce the hardware complexity of HOG. In [6] and [7], the computational complexity of HOG is reduced with cell-based scanning and simultaneous SVM calculation using FPGA and ASIC implementations for full HD resolution; however, the implementations use the parameters as suggested in [4]. Various hardware optimizations are presented in [15–22] for an efficient pedestrian detection system. However, for real-time PD with power and area constraints, it is imminent to find the set of parameters of HOG that provide the best compromise in terms of computational complexity and accuracy. Recently, a hardware architecture for fixed point HOG implementation has been presented [8] where the bit-width has been optimized to achieve significant improvement in power and throughput. We believe that in addition to bit-width there are other parameters which need to be optimized to provide a holistic understanding of the relationship between accuracy, speed, power, and complexity. Moreover, sliding window based pedestrian detection requires detection to be performed at multiple scales of image. It has been shown that the best detection performance can be achieved with scale factor (the ratio to scale the image after each detection) of at-least 1.09. This results in processing around 45 scales for full-HD resolution. As shown in Fig. 1, a combination of the number of scales (=45) required for maximum accuracy [8, 9] and throughput for real-time pedestrian detection at full-HD resolution has not been achieved before.

The key contributions of our work can be summarized as follows.

- We present parallel implementation of various blocks of HOG-based PD on an FPGA. Parallel implementation has been used to improve the speed of PD.

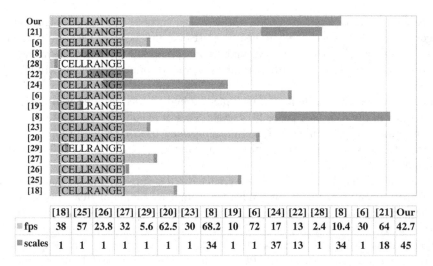

	[18]	[25]	[26]	[27]	[29]	[20]	[23]	[8]	[19]	[6]	[24]	[22]	[28]	[8]	[6]	[21]	Our
▪ fps	38	57	23.8	32	5.6	62.5	30	68.2	10	72	17	13	2.4	10.4	30	64	42.7
▪ scales	1	1	1	1	1	1	1	34	1	1	37	13	1	34	1	18	45

Fig. 1. Comparison of the state of the art in terms of frames per second and number of scales.

- We derive the accuracy, speed, power, and hardware complexity results of HOG-based PD with different choices of cell sizes and number of histogram bins.
- We show that by using the right choice of cell size and number of histogram bins, a significant reduction in power consumption and increase in throughput can be achieved with reasonable accuracy.
- Finally features are refined by removing insignificant gradients which results in not only improvement of throughput and power consumption but also accuracy.

This chapter is an extension of our original paper [10] including more detailed literature survey and hardware implementation details. The rest of the chapter is organized as follows. Section 2 summarizes the state of the art in PD. In Sect. 3, a brief overview of HOG is presented. The proposed hardware implementation is discussed in Sect. 4. In Sect. 5, the accuracy, speed, power, and hardware complexity results are shown for different choices of parameters and the optimal choice of parameters under given constraints is described. Section 6 concludes the work.

2 Literature Survey

Numerous efforts have been made in the past to perform PD efficiently. An extensive survey of PD schemes is given in [9]. Generally, these approaches can be classified into two categories: segmentation-based approaches [34] and sliding window-based approaches [11]. A segmentation-based approach processes the whole frame at once and extracts segments of the frame which include pedestrians. On the other hand, sliding window-based approaches divide a frame into multiple, overlapping windows and search pedestrians in each of these windows.

2.1   Sliding Window Based Pedestrian Detection

Recently, researchers have put more effort into sliding window approach as this approach simplifies the problem of PD to binary classification in a given window. Sliding window-based approaches can be further subdivided into rigid and part-based methods. The rigid schemes consider the window holistically to identify a pedestrian. Part-based schemes, such as [35, 36] identify different parts of a pedestrian in a window, and decide the presence of a pedestrian based on the location and confidence (accuracy) of detected parts. Part-based schemes have been shown to perform better compared to rigid schemes as the decision in these methods is based on the aggregate of decisions for different parts and these schemes can handle occlusion better compared to rigid schemes. However, the higher computational complexity of part-based methods makes them infeasible for real-time applications. Rigid schemes utilize a single feature or multiple features to detect an object. We have categorized the schemes depending on the feature and implementation platform.

Single Feature Pedestrian Detection. In [11], the authors use Haar-like features with Support Vector Machines (SVM) to identify objects in a scene. Their method was advanced in [37] for face detection, where the authors obtained an astounding increase in speed by using integral images to compute Haar-like features. Furthermore, cascaded boosted trees were used for classification. The method of [37] was used for PD in [38]. However, using Haar-like features for PD detection did not show much promise until recently [4] due to their low accuracy.

A set of rich and compact features was required to improve PD. Rich features were needed to extract maximum information from a window and compactness was needed to better generalize from training to testing. HOG [4] performed both these tasks by including the complete (or rich) gradient information of a window into compact histograms. They trained Linear Support Vector Machine (LinSVM) framework for classification. Furthermore, they developed and used the INRIA pedestrian dataset, which was the most extensive dataset for PD at that time. Resultantly, their method achieved significant improvement in accuracy of PD compared to the previous schemes.

Since its inception, HOG has influenced most of the modern PD methods. In [39], gradients in local patches, similar to HOG, are used to represent shape descriptors. These shape descriptors are combined to create a single feature which is classified using boosted trees. The method is used for PD as well as detection of other objects. Edgelets, used in [40] and [41], have been used to learn and classify body parts with boosted trees. Other variations include distance transform and template hierarchy [42] to match images with templates, granularity-tunable gradients partition to define spatial and angular uncertainty of line segments [43] and its extension to spatiotemporal domain [44], shape features [45] and finally motion based features [46, 47].

Multiple Feature Pedestrian Detection. To further enhance the PD accuracy of HOG, researchers have complimented HOG with other features. Local Binary Pattern (LBP) is a very simple feature based on magnitude comparison of surrounding pixels, and has typically been used for texture classification [48] and face detection [49]. It has also been used in PD [50]. In [51], the authors present a feature combining both HOG and LBP and use linSVM for classification. They show that this combination improves the PD performance under partial occlusion. In [52], the authors use implicit segmentation and divide the image into separate foreground and background, followed by HOG. HOG, LBP and local ternary patterns were combined in [53] for pedestrian and object detection. Gradients information and HOG, textures (co-occurrence matrices), and color frequency are combined in [54]. Partial least squares are used to reduce the dimensions of the feature and SVM is used for classification. HOG has been combined with Haar-like [55], shapelets [39], color self-similarity and motion [56] features as well. Note that none of these features when used independently from HOG has been able to outperform HOG.

2.2 Real Time Pedestrian Detection

Numerous applications require PD at fast rates. For such applications, it is more suitable to use task-specific (GPU, FPGA, ASIC) rather than general-purpose platforms. A fine grain parallel ASIC implementation of HOG-based PD is presented in [7]. In [15], simplified methods are presented for division and square root operations for use in HOG. However, by employing their methods, the accuracy of PD is severely degraded. A multiprocessor system on chip (SoC) based hardware accelerator for HOG feature extraction is described in [16]. In [17], the authors reuse the features in blocks to construct the HOG features of overlapping regions in detection windows and then use interpolation to efficiently compute the HOG features for each window.

In [18], the authors developed an efficient FPGA implementation of HOG to detect traffic signs. In [19], a real-time PD framework is presented which utilizes an FPGA for feature extraction and a GPU for classification. A deep-pipelined single chip FPGA implementation of PD using binary HOG with decision tree classifiers is discussed in [20]. A heterogeneous system is presented in [22] to optimize the power, speed and accuracy.

From the discussion above, we notice that HOG is integral to most PD algorithms. Efforts have been made in the research community to either improve the accuracy of PD or reduce the hardware complexity of HOG. In our study, we are yet to find an effort which analyzes the effects of reducing hardware costs on accuracy of HOG. For real-time PD with power and area constraints, it is imminent to find the set of parameters of HOG that provide the best compromise in terms of computational complexity and accuracy.

3 Overview of HOG

In this section, we present a brief overview of the HOG algorithm for PD. Although HOG can be used in a part-based PD scheme, we limit our discussion to the rigid HOG as described in the original paper [4]. A block diagram showing functional blocks of the algorithm is shown in Fig. 2.

In HOG, a search window is divided into multiple overlapping blocks which are further divided into cells as shown in Fig. 3, where $(w_F, h_F), (w_W, h_W), (w_B, h_B),$ $(w_C, h_C),$ are the frame, window, block and cell (width, height) respectively. N_{bin} is the

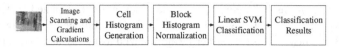

Fig. 2. Block diagram of HOG based pedestrian detection

number of histogram bins. The blocks have an overlap of 50 %, creating a dense grid over the search window. So a single $(w_W \times h_W)$ window has $n_C = w_W/w_C \times h_W/h_C$ cells and $n_B = \left(\frac{w_W - w_C}{w_C} \times \frac{h_W - h_C}{h_C} \right)$ blocks. Gradient features are extracted from these blocks and cells, and are concatenated to create a single feature vector for the whole window.

A filter with coefficients $[-1, 0, 1]$ is applied to the window in horizontal and vertical directions, creating the images G_x and G_y, respectively. These images are used to generate the gradient magnitude image, G_m, and the gradient orientation image, G_o, for each pixel (x, y) as follows.

$$G_m(x, y) = |G(x, y)| = \sqrt{G_x(x, y)^2 + G_y(x, y)^2} \tag{1}$$

$$G_o(x, y) = \tan^{-1} \frac{G_y(x, y)}{G_x(x, y)} \tag{2}$$

The histogram used in the feature accumulates the orientation information of an image. Each histogram has multiple bins, where each bin represents a specific orientation in the interval $[0, \pi)$. The value $G_m(x, y)$ is added to the bin of the histogram which corresponds to $G_o(x, y)$. Such histograms are developed for every cell of the window, as shown in Fig. 3.

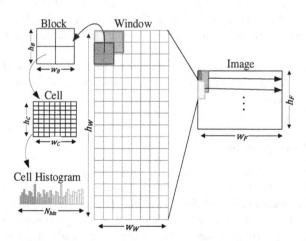

Fig. 3. A depiction of image division for sliding window based object detection. An input image $(w_F \times h_F)$ is divided into overlapping windows. The window is divided into overlapping blocks which are further divided into cells. A histogram is generated for every cell.

The cell histograms belonging to a single block are concatenated to form block histogram of length $M = 4 \times N_{bin}$, where, N_{bin} is the number of bins in each cell histogram. Block histograms are further L2-normalized using (3), and then added to the feature vector. *L2-norm* for an un-normalized feature vector v, is given by,

$$x_i^b = \frac{v_i^b}{\sqrt{\|v\|_2^2 + \varepsilon^2}}. \tag{3}$$

where $i = 1, \ldots, M, b = 1, \ldots, n_B, \|v\|_2^2 = v_1^2 + v_2^2 + \ldots + v_M^2$ and ε is a small constant to avoid division by zero. L2-normalization is performed to improve robustness against illumination changes.

For classification, LinSVM is used. From an implementation perspective, a weight vector is obtained after the training stage. During classification, a dot product of the feature extracted from the window and the weight vector is compared against a threshold. If the dot product is greater than the threshold, then a pedestrian is identified.

4 Hardware Architecture

The hardware implementation of HOG presents a unique challenge, which is quite distinct from the software implementation. First, we cannot store and access a complete frame, and read and write from multiple addresses at once as this will require unrealistically large hardware resources. Second, floating point operations are quite costly in hardware, as they use more FPGA area and runs at a lower frequency; therefore, we avoid them in hardware implementation. Finally, the choice of parameters affects hardware complexity significantly compared to software implementation.

Our key objectives in this implementation are to maintain the maximum accuracy and minimum power consumption while performing real time PD by controlling local features. Hardware/memory optimization is done using optimal values of these features. The optimized architecture thus obtained results in a reduced workload and low bandwidth.

The conceptual block diagram of the proposed HOG Accelerator (HOG-Acc) is shown in Fig. 4. In the following we present a description of the major functional blocks shown in Fig. 4.

4.1 Gradient Computation

To compute the gradient magnitude and orientation, the horizontal and vertical gradient images, i.e., G_x and G_y, need to be generated. Gradient is computed over the 3×3 neighborhood of each pixel; therefore, two line buffers are required to store two consecutive scan lines of the image to maintain a 3×3 neighborhood of every pixel.

A straight forward computation of the gradient magnitude, as given in (1), will require the implementation of the square root operation, which will consume significant hardware resources; thereby, delay and power consumption will increase. In order to

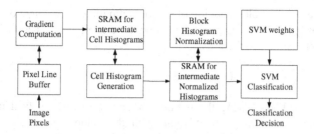

Fig. 4. Block diagram of hardware architecture. Gradient is computed over input pixels stored in Pixel Line buffer, cell histograms are then built using gradients, SRAM is used to store intermediate cell histograms. Next steps are normalization of histograms generated and finally classification.

reduce the computational complexity, the following approximations from [15] have been used to compute the gradient magnitude and orientation.

$$G_M(x, y) \approx \max((0.875a + 0.5b), a), \tag{4}$$

where,

$$a = \max\big(G_x(x, y), G_y(x, y)\big), \tag{5}$$

and

$$b = \min\big(G_x(x, y), G_y(x, y)\big). \tag{6}$$

The circuitry for gradient magnitude computation is shown in Fig. 5. Equations (5) and (6) are implemented using a single compare operation, while (4) requires four shift operations yielding $0.875a$ and $0.5b$, then an adder and one more comparator is used to give the final gradient magnitude.

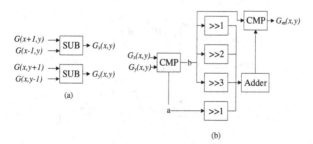

Fig. 5. Gradient magnitude computation module. (a) Simply subtract the horizontal and vertical neighboring pixels to compute the horizontal and vertical gradients. (b) Gradient magnitude is computed by shift and compare operations to implement (4)–(6)

Similarly, a direct implementation of (2) for computing the gradient orientation will require two costly hardware operations: the inverse tangent and division. To reduce the complexity, (2) can be rewritten as

$$G_x(x, y) \tan(G_o(x, y)) = G_y(x, y). \tag{7}$$

The problem of identifying the gradient orientation can be solved using (7) as: multiplying the horizontal gradient value with the values of the right column of Table 1; the product which best matches against the vertical gradient indicates the gradient of the pixel. Note that even the multiplication operation is not required, as the product with the values in the right column of Table 1 can be performed by simple arithmetic shifting.

Table 1. Approximated values of $tan\theta$

Tangent	Approximated value
$tan0°$	0
$tan10°$	$2^{-3} + 2^{-4}$
$tan20°$	$2^{-2} + 2^{-3}$
$tan30°$	$2^{-1} + 2^{-4} + 2^{-6}$
$tan40°$	$2^{-1} + 2^{-2} + 2^{-4}$

The circuit to compute the histogram bin is shown in Fig. 6. It consists of two parts, one deals with the quadrant decision and other decides the bin. Comparing horizontal and vertical neighboring pixels sets the Q-flag value which indicates whether the bin lies in first or second quadrant. Once we know the quadrant, we have to decide which histogram bin the orientation value lies in. By using Table 1 to approximate the value of tangent function at different orientations, complex operations such as inverse tangent and division can be avoided. The hardware utilizes only comparators, shifters and adders, hence reducing the complexity significantly.

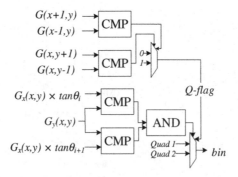

Fig. 6. Bin computation module: bin quadrant is decided using a Q-flag, which is computed by comparing horizontal and vertical pixels, histogram bin is then decided implementing (7) using comparators and AND gate.

It has been shown in [8] that bit-width assigned to magnitude has a significant impact on accuracy, throughput and power consumption as it affects the data sizes at all the next stages. In [8] fixed-point implementation is considered and bit width of gradient magnitude is optimized as 13 bits (9:4 (integer: fractional)). We argue that using only integer values of gradient magnitude can further improve the accuracy, throughput and power consumption. The details are given in Sect. 4. The key insight is that by using integer values for gradient magnitudes, we can remove the histogram values which are less significant. The advantages are twofold. (1) It reduces the hardware complexity due to reduced bit width and integer operations. (2) It improves the accuracy because removing these petty gradient magnitudes enhances the feature vector for training and classification.

4.2   Cell Histogram Generation

We propose a parallel Cell Histogram Generation (CHG) module as shown in Fig. 7. Gradient magnitudes and orientation bins for every $w_C \times h_C$ pixels are given as input to CHG. Decoders and adders are used to build the histogram. Each bin value is given as input to the decoder. Only one output is set to '1' corresponding to the specific bin; gradient magnitude for that bin hence propagates to the input of adder, where all the magnitudes of the same bin are added.

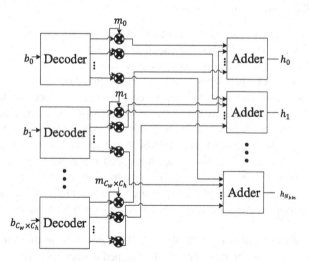

Fig. 7. Cell Histogram Generation (CHG) Engine: Histogram bins and gradient magnitudes are given as input and cell histograms are generated.

The decoder size is dependent on the number of bits required to represent single bin, i.e. if number of histogram bins increase the size of decoder increases. On the other hand, the cell size ($C_{size} = w_C \times h_C$) affects the number of decoders as the total number of decoders required equals C_{size}. Multipliers required for CHG are dependent on both

N_{bin} and C_{size}. Multipliers in each stage depend on N_{bin} while number of stages depends on C_{size}. Finally, the number of adders is equivalent to the N_{bin} chosen. Adder size, however, varies according to C_{size}. We can clearly see that the complexity of CHG is strongly dependent on N_{bin} and C_{size}.

Since pixels are coming row by row, we have to maintain cell histograms for multiple blocks and windows as each row has multiple windows. Therefore, the gradient magnitudes and orientations computed for every w_C pixels (one cell) are concatenated and stored in memory. Pixels of row index which is a multiple of h_C indicate the completeness of cell. This row is directly stored into registers. At the start of every such row, respective values of previous rows for the particular cell are read into registers from block RAM every clock cycle. As we have considered $w_C = h_C$ the cell completes in horizontal and vertical directions simultaneously. Hence, the number of shift registers required is equivalent to h_C. Each shift register stores magnitudes and bins for w_C pixels. After w_C cycles the data of one cell is completed so it is shifted to the memory, which in turn writes the data for the previous row in the next register.

The resultant cell histogram is given to the next stage for processing. This is done every time the new cell is completed. i.e. when the row index is a multiple of h_C and column index is a multiple of w_C. The cell histograms for multiple windows in a frame are stored in memory while they are shifted to registers for each active window (the window whose cells histograms are completed).

4.3 Block Histogram Normalization

Cell histograms are maintained in the memory till four neighboring cells are completed and a block is obtained. Note that the memory required to store the cell histograms increases with smaller cells (more cells per row and column) and larger number of histogram bins for every cell (more data per cell). In other words, C_{size} affects the memory locations required while N_{bin} influences the width of each location.

The histogram is normalized using the Block Histogram Normalization Engine (BHN) shown in Fig. 8. Normalization is performed every time a new block is completed. Each histogram value in a block is squared and added. The sum is given as input to inverse square root module which is approximated using "fast inverse square root" algorithm [30]. In summary, logical shifting, subtraction and finally one iteration of Newton's method approximates the inverse square root. Finally, the result of inverse square root is multiplied with each histogram value to generate the normalized block histogram.

It is seen that the number of multipliers in BHN depends on the size of the block histogram, which is related to number of bins assigned to each cell histogram. Adding a single bin to cell histogram adds eight multipliers to the hardware. The adder size also increases proportionately.

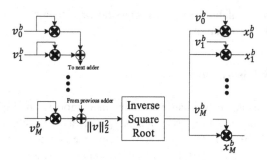

Fig. 8. Block Histogram Normalization (BHN) engine: un-normalized Block histograms (concatenated cell histograms) are used as inputs to generate normalized block histograms.

4.4 SVM Classification

The normalized histograms obtained from the BHN block are again stored in the memory. Once normalized histograms for the whole window are available classification can be performed which can consume a fair amount of memory. Performing classification for the whole window at once also requires a large number of multipliers and adders. The situation gets worse as the feature vector size increases with smaller cell sizes or large number of bins. Therefore, we have opted for partial classification by dividing the classification for the whole window into multiple stages. The hardware shown in Fig. 9. is reused at every stage. The strategy behind reusing the hardware is very straightforward. Since it takes w_C cycles to completely process a cell, we have reused the same hardware over these w_C cycles doing partial classification every N_B/w_C blocks. So the number of partial classification stages is equal to C_w. The results of each stage are accumulated to get the final classification result.

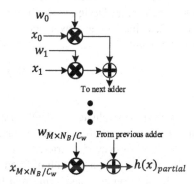

Fig. 9. Partial Classification Engine (PCE), single stage of LinSVM classification to be performed for whole window. Inputs are normalized histograms while output is the partial classification result.

The key observation is that the cell size effects the hardware complexity in two ways. First, it has a direct impact on feature vector size. Second, larger the cell size, more cycles will be available to perform classification, thereby, smaller hardware is required for partial classification.

5 Results and Discussion

In this section, we evaluate our hardware implementation for multiple cell sizes and histogram bins to obtain optimal set of these parameters. Results are presented for full-HD (1920 × 1080) resolution videos. Window size is considered to be 64 × 128. Block size is 2 × 2 cells, while block and window step size is one cell for both horizontal and vertical directions. Scale factor to rescale images is set to 1.05. This results in 45 scales to be processed per frame. Other parameters depend on the choice of cell size and histogram bins.

Here, we first present our experimental setup then we analyze the effect of different cell sizes and histogram bins on accuracy, throughput and power. Using these results, parameters yielding least power and maximum throughput with negligible loss in accuracy are selected. Finally, using these parameters comparison with the state of the art object detection implementations is presented.

5.1 Experimental Setup

We have implemented our system on Xilinx Virtex 7 (XC7VX485T) FPGA. There are 75,900 slices, 607,200 Configurable Logic Blocks (CLBs) and 485,760 logic cells in this FPGA. Moreover, 37,080 Kb block RAM and 2,800 DSP slices are present. Image rescaling and window sampling is done for positive and negative images and then sent to HOG-Acc for processing which returns the detection result. Processing 45 scales requires a large amount of memory and pipelined stages so we have utilized the time multiplexing approach of [21]. The host software is written using Visual Studio 2012 and Verilog is used for HOG-Acc design. Design is synthesized using Xilinx ISE 14.7 and along with Modelsim 10.2, a hardware/software co-simulation is performed to verify the implementation functionality.

5.2 Accuracy Analysis

We have used INRIA dataset [31], to evaluate our HOG implementation. There are several other datasets available for pedestrian detection evaluation like Caltech [32], ETH [33], and Daimler [3]. We have, however, restricted our results to INRIA because it provides us with a reasonable variety of images with different poses and backgrounds so these results can be generalized to other datasets and real life scenarios.

All detection results are collected, and afterwards recall is calculated from number of true positives (TP) and false negatives (FN) as shown in (8).

$$Recall = \frac{TP}{TP + FN} \tag{8}$$

A false positive per window (FPPW) of 10^{-4} is mostly considered in literature for pedestrian detection results. We also present the *Miss Rate (1-Recall)* results for $FPPW = 10^{-4}$ for multiple cell sizes and number of bins. The results are shown in Fig. 10. It is seen that generally larger histogram bins gives better detection rates. This is obvious, as more histogram bins allow fine division gradient orientations, hence better feature vector for training and classification. On the other hand, improvement is seen in detection rates by increasing cell size up to a certain value and it drops increasing cell sizes too much. Smaller cell sizes provide a dense grid of blocks and windows in a frame, therefore, using smaller cells would improve accuracy. However, using too small cell sizes results in degraded performance because there are not enough distinguishing features within the cells. Minimum miss rate of 12 % is achieved at $(C_{size}, N_{bin}) = (7 \times 7, 11)$.

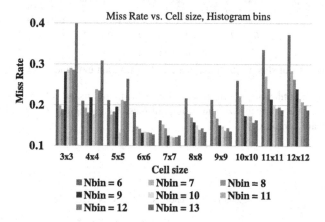

Fig. 10. Accuracy analysis, miss rate generally reduces with increasing cell sizes and decreasing number of bins

5.3 Throughput and Power Consumption Analysis

Power consumption and throughput are directly related to the hardware resources used. In the previous section it is seen that the cell size and histogram bins has significant impact on hardware complexity. The effect on different hardware components for different cell sizes and number of bins for a single core is shown in Fig. 11. We see a significant reduction in hardware resources by increasing cell size or reducing number of bins. The reasons being discussed in previous section for independent blocks.

Number of frames processed per second (fps) is dependent on the maximum frequency at which the hardware can operate. In our hardware architecture it is mainly dependent on the size of partial classification engine and the block normalization engine. As discussed in previous section, the complexity of PCE is heavily dependent

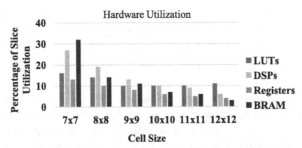

(a) Percentage of slice utilization for multiple cell sizes and fixed N_{bin}=9.

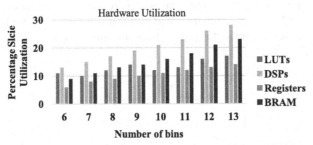

(b) Percentage of slice utilization for multiple histogram bins and fixed C_{size}=8x8

Fig. 11. Hardware utilization comparison. Breakdown of usage of multiple slices of Xilinx FPGA with varying cell size and histogram bins. Increasing any one of them results in increased hardware complexity.

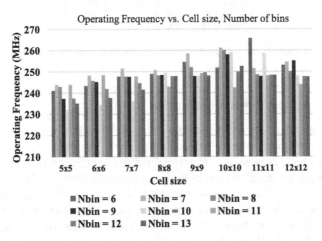

Fig. 12. Throughput analysis, an increase in throughput is seen for bigger cell sizes and histogram bins.

on C_{size} while that of BHN depends on both C_{size} and N_{bin}. Figure 12 shows the results. We get the maximum frequency at the point where both PCE and BHN have overall minimal hardware complexity. Specifically, the $(C_{size}, N_{bin}) = (11 \times 11, 6)$.

We have used Xilinx Xpower analyzer (XPA) 14.7 to estimate the deviations in power consumption by varying the parameters. We have simulated the hardware and created 'Value Change Dump' (vcd) files are used to set the toggle rates of all signals. Post place and route results are obtained and are shown in Fig. 13. Power consumption increases by reducing the cell size or increasing the histogram bins. This is fairly understandable due to the fact that both these parameters increase the hardware complexity due to increase in the feature vector size. Minimum power consumption is 9.98 W with $(C_{size}, N_{bin}) = (12 \times 12, 6)$, the maximum cell size and minimum histogram bins as expected.

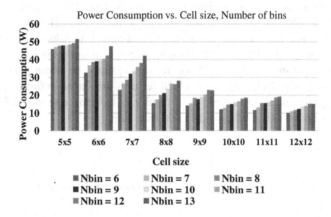

Fig. 13. Power consumption analysis, large cell size and smaller number of histogram bins results in low power consumption.

5.4   Choice of Parameters

We have seen from the previous analysis that there does not exist a set of parameters which give us best accuracy, power and throughput. Improving the accuracy worsens the power and throughput while maintaining minimum power and maximum throughput severely degrades the accuracy. Similarly, trying to improve throughput may degrade power consumption significantly and vice versa. However, accuracy is changing very slightly at certain regions in Fig. 9. Similarly, there are more than one sets of parameters which give almost the same power consumption. This allows us to select the best of one of these metrics while slightly compromising on another metric. We can achieve best results by selecting $(C_{size}, N_{bin}) = (9 \times 9, 10)$. Further we obtained results for miss rate by changing bit-width for this optimal parameter set. The results are shown in Fig. 14. Bit-width is hence set to eight bits as it gives maximum accuracy and minimum hardware complexity. Note that this further results in reduced bit-width in all the next blocks.

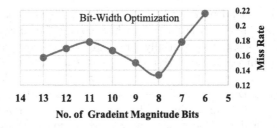

Fig. 14. Variation in miss rate for $(C_{size}, N_{bin}) = (9 \times 9, 10)$.

Parameters optimized for low power and high speed are shown in Tables 2 and 3 comparing the throughput and energy consumption results with the other state of the art FPGA and GPU implementations. We have presented three results. (1) HOG_{CONV}, which shows the results for conventionally used parameters. We can achieve 32 fps for full-HD while dissipating 0.656 J/frame and 15 % miss rate. (2) HOG_{OCB}, presents results for optimized cell size and histogram bins. Frame rate achieved by optimizing cell size and histogram bins is 39.2 fps with energy consumption of 0.484 J/frame while maintaining a miss rate at 15 %. Gradient magnitude bit-width is considered to be 13 bits. (3) Finally, HOG_{OCB-RF}, in which features are further refined by removing insignificant gradients, is presented. This results in a frame rate of 42.7 fps while energy consumption is 0.45 J/frame at 13 % miss rate.

Table 2. Comparison of parameters and throughput for various GPU and FPGA implementations

	Cell size	Histogram bins	Win. stride	# scales	Resolution	Windows/frame	FPS
GPU implementation							
[22, 24]	8 × 8	9	8	37	1024 × 768	–	17
[25]	8 × 8	–	–	>1	640 × 480	4,096	57
[26]	8 × 8	8	2	>1	640 × 480	50,000	23.8
[27]	8 × 8	9	8	1	640 × 480	–	32
[28]	8 × 8	9	4	>1	1280 × 960	150,000	2.4
[29]	8 × 8	9	–	>1	640 × 480	–	5.6
FPGA implementation							
[18]	8 × 8	8	4	>1	320 × 240	3,615	38
[19]	8 × 8	9	–	1	800 × 600	1000	>10
[20]	8 × 8	8	9	1	640 × 480	1,540	62.5
[21]	8 × 8	9	8	18	1920 × 1080	27,960	64 (estimated)
[22]	8 × 8	9	8	13	1024 × 768	20,868	13
[23]	8 × 8	8	4	>1	640 × 480	56,466	30 (estimated)
[6]	8 × 8	9	8	1	800 × 600	5,580	72
[8]	8 × 8	9	4	34	640 × 480	121,210	68.18
[8]	8 × 8	9	4	34	1600 × 1200	1,049,886	10.41 (estimated)
HOG_{CONV}	8 × 8	9	8	45	1920 × 1080	264,062	32
HOG_{OCB}	9 × 9	10	8	45	1920 × 1080	264,062	39.2
HOG_{OCB-RF}	9 × 9	10	8	45	1920 × 1080	264,062	42.7

Table 3. Comparison of parameters and energy consumption for various GPU and FPGA implementations

	Cell size	Histogram bins	Win. stride	# scales	Resolution	Windows/frame	Power (W)	Energy (J/Frame)
[8] (GPU)	8 × 8	9	4	34	640 × 480	121,210	225	17
[8] (FPGA)	8 × 8	9	4	34	640 × 480	121,210	37	0.54
HOG_{CONV}	8 × 8	9	8	45	1920 × 1080	264,062	21	0.656
HOG_{OCB}	9 × 9	10	8	45	1920 × 1080	264,062	19	0.485
$HOG_{OCB\text{-}RF}$	9 × 9	10	8	45	1920 × 1080	264,062	19.276	0.451

6 Conclusion

We have presented fully parallel architectures for various modules of pedestrian detection system utilizing Histogram of oriented gradients (HOG). HOG has shown high detection accuracy but the detection speed and power consumption are major bottlenecks for real time embedded applications. We have optimized parameters, cell size and histogram bins, to achieve low power and high throughput while maintaining the detection accuracy. Feature refinement is done to further improve the results.

Combination of optimal parameters and our hardware accelerator results in a frame rate of 42.7 fps for full-HD resolution and lowers the energy consumption by 97.34 % and 16.4 % while improving the accuracy by 2 % as compared to state of the art GPU and FPGA implementations respectively. This work can be extended to use multiple cores on a single FPGA or using multiple FPGAs to further increase throughput while an ASIC implementation would greatly reduce the power consumption. It can also be extended to include other features and classifiers or combinations of those to optimize for objects other than pedestrians.

Acknowledgments. This work was supported by the Center of Integrated Smart Sensors funded by Ministry of Science, ICT & Future Planning as Global Frontier Project (CISS-2013M3 A6A6073718).

References

1. Shankar, U.: Pedestrian roadway fatalities. Department of Transportation Technical report (2003)
2. Geronimo, D., Lopez, A.M., Sappa, A.D., Graf, T.: Survey on pedestrian detection for advanced driver assistance systems. IEEE Trans. Pattern Anal. Mach. Intell. **32**(7), 1239–1258 (2010). 1, 2, 10, 16, 18
3. Ess, A., Leibe, B., Schindler, K., Van Gool, L.: A mobile vision system for robust multi-person tracking. In: IEEE Conference on Computer Vision and Pattern Recognition (CVPR), pp. 1–8 (2008)
4. Dalal, N., Triggs, B.: Histograms of oriented gradients for human detection. In: Proceedings of IEEE Conference on Computer Vision Pattern Recognition, vol. 1, pp. 886–893, June 2005
5. Lowe, D.G.: Distinctive image features from scale-invariant keypoints. IJCV **60**(2), 91–110 (2004)

6. Mizuno, K., Terachi, Y., Takagi, K.: Architectural study of HOG feature extraction processor for real-time object detection. In: Proceedings of IEEE Workshop Signal Processing Systems, pp. 197–202, October 2012
7. Takagi, K., et al.: A sub-100-milliwatt dual-core HOG accelerator VLSI for real-time multiple object detection. In: ICASSP (2013)
8. Ma, X., Najjar, W.A., Roy-Chowdhury, A.K.: Evaluation and acceleration of high-throughput fixed-point object detection on FPGAs. IEEE Trans. Circ. Syst. Video Technol. **25**(6), 1051–1062 (2015)
9. Dollar, P., Wojek, C., Schiele, B., Perona, P.: Pedestrian detection: an evaluation of the state of the art. IEEE Trans. Pattern Anal. Mach. Intell. **34**(4), 743–761 (2012)
10. Khan, A., Khan, M.U.K., Bilal, M., Kyung, C.-M.: Hardware architecture and optimization of sliding window based pedestrian detection on FPGA for high resolution images by varying local features. In: VLSI-SoC 2015, pp. 142–148 (2015)
11. Papageorgiou, C., Poggio, T.: A trainable system for object detection. Int. J. Comput. Vis. **38**(1), 15–33 (2000)
12. Oren, M., Papageorgiou, C., Sinha, P., Osuna, E., Poggio, T.: Pedestrian detection using wavelet templates. In: Proceedings of IEEE Conference on Computer Vision Pattern Recognition, pp. 193–199, June 1997
13. Lowe, G.: Distinctive image features from scale-invariant keypoints. Int. J. Comput. Vis. **60** (2), 91–110 (2004)
14. Cheng, H., Zheng, N., Qin, J.: Pedestrian detection using sparse Gabor filter and support vector machine. In: Proceedings of IEEE Intelligent Vehicles Symposium, pp. 583–587, June 2005
15. Chen, P.Y., Huang, C.C., Lien, C.Y., Tsai, Y.H.: An efficient hardware implementation of HOG feature extraction for human detection. IEEE Trans. Intell. Transp. Syst. **15**(2), 656–662 (2014)
16. Lee, S.E., Min, K., Suh, T.: Accelerating histograms of oriented gradients descriptor extraction for pedestrian recognition. Comput. Elect. Eng. **39**(4), 1043–1048 (2013)
17. Pang, Y., Yuan, Y., Li, X., Pan, J.: Efficient HOG human detection. Signal Process. **91**(4), 773–781 (2011)
18. Hiromoto, M., Miyamoto, R.: Hardware architecture for high-accuracy real-time pedestrian detection with CoHOG features. In: Proceedings of IEEE ICCVW, pp. 894–899 (2009)
19. Bauer, S., Kohler, S., Doll, K., Brunsmann, U.: FPGA-GPU architecture for kernel SVM pedestrian detection. In: 2010 IEEE Computer Society Conference on Computer Vision and Pattern Recognition Workshops (CVPRW), pp. 61–68, June 2010
20. Negi, K., Dohi, K., Shibata, Y., Oguri, K.: Deep pipelined one-chip FPGA implementation of a real-time image-based human detection algorithm. In: Proceedings of International Conference on FPT, pp. 1–8, 12–14 December 2011
21. Hahnle, M., Saxen, F., Hisung, M., Brunsmann, U., Doll, K.: FPGA based real-time pedestrian detection on high-resolution images. In: 2013 IEEE Conference on Computer Vision and Pattern Recognition Workshops (CVPRW), pp. 629–635, June 2013
22. Blair, C., Robertson, N., Hume, D.: Characterizing a heterogeneous system for person detection in video using histograms of oriented gradients: power versus speed versus accuracy. IEEE J. Emerg. Sel. Top. Circ. Syst. **3**(2), 236–247 (2013)
23. Kadota, R., Sugano, H., Hiromoto, M., Ochi, H., Miyamoto, R., Nakamura, Y.: Hardware architecture for HOG feature extraction. In: 5th International Conference on Intelligent Information Hiding and Multimedia Signal Processing (IIHMSP), pp. 1330–1333 (2009)
24. OpenCV: http://opencv.org/

25. Sudowe, P., Leibe, B.: Efficient use of geometric constraints for sliding-window object detection in video. In: Crowley, J.L., Draper, B.A., Thonnat, M. (eds.) ICVS 2011. LNCS, vol. 6962, pp. 11–20. Springer, Heidelberg (2011)
26. Machida, T., Naito, T.: GPU & CPU cooperative accelerated pedestrian and vehicle detection. In: IEEE International Conference on Computer Vision Workshops (ICCV Workshops), pp. 506–513 (2011)
27. Yan-ping, C., Shao-zi, L., Xian-ming, L.: Fast HOG feature computation based on CUDA. IEEE Int. Conf. Comput. Sci. Autom. Eng. (CSAE) **4**, 748–751 (2011)
28. Bilgic, B., Horn, B.K.P., Masaki, I.: Fast human detection with cascaded ensembles on the GPU. In: 2010 IEEE Intelligent Vehicles Symposium (IV), pp. 325–332 (2010)
29. Prisacariu, V., Reid, I.: fastHOG - a real-time GPU implementation of HOG. Department of Engineering Science, Oxford University, Technical report 2310/09 (2009)
30. http://en.wikipedia.org/wiki/Fast_inverse_square_root
31. INRIA Person Dataset. http://pascal.inrialpes.fr/data/human/
32. Dollar, P., Wojek, C., Schiele, B., Perona, P.: Pedestrian detection: a benchmark. In: IEEE Conference on Computer Vision and Pattern Recognition (CVPR), pp. 304–311 (2009)
33. Enzweiler, M., Gavrila, D.: Monocular pedestrian detection: survey and experiments. IEEE Trans. Pattern Anal. Mach. Intell. **31**(12), 2179–2195 (2009)
34. Gu, C., Lim, J.J., Arbelaez, P., Malik, J.: Recognition using regions. In: IEEE Conference on Computer Vision and Pattern Recognition (2009)
35. Felzenszwalb, P., McAllester, D., Ramanan, D.: A discriminatively trained, multiscale, deformable part model. In: IEEE Conference on Computer Vision and Pattern Recognition (2008)
36. Felzenszwalb, P.F., Girshick, R.B., McAllester, D., Ramanan, D.: Object detection with discriminatively trained part based models. IEEE Trans. Pattern Anal. Mach. Intell. **99** (2009). PrePrints
37. Porikli, F.M.: Integral histogram: a fast way to extract histograms in Cartesian spaces. In: IEEE Conference on Computer Vision and Pattern Recognition (2005)
38. Viola, P.A., Jones, M.J., Snow, D.: Detecting pedestrians using patterns of motion and appearance. Int. J. Comput. Vis. **63**(2), 153–161 (2005)
39. Sabzmeydani, P., Mori, G.: Detecting pedestrians by learning shapelet features. In: IEEE Conference on Computer Vision and Pattern Recognition (2007)
40. Wu, B., Nevatia, R.: Detection of multiple, partially occluded humans in a single image by Bayesian combination of edgelet part det. IEEE Int. Conf. Comput. Vis. **6**(10), 11 (2005)
41. Wu, B., Nevatia, R.: Cluster boosted tree classifier for multi-view, multi-pose object detection. In: ICCV (2007)
42. Gavrila, D.M.: A Bayesian, exemplar-based approach to hierarchical shape matching. IEEE Trans. Pattern Anal. Mach. Intell. **29**(8), 1408–1421 (2007)
43. Liu, Y., Shan, S., Zhang, W., Chen, X., Gao, W.: Granularity tunable gradients partition descriptors for human detection. In: IEEE Conference on Computer Vision and Pattern Recognition (2009)
44. Liu, Y., Shan, S., Chen, X., Heikkila, J., Gao, W., Pietikainen, M.: Spatial-temporal granularity-tunable gradients partition (STGGP) descriptors for human detection. In: Daniilidis, K., Maragos, P., Paragios, N. (eds.) ECCV 2010, Part I. LNCS, vol. 6311, pp. 327–340. Springer, Heidelberg (2010)
45. Gavrila, D.M., Philomin, V.: Real-time object detection for smart vehicles. In: IEEE International Conference on Computer Vision, pp. 87–93 (1999)
46. Dalal, N., Triggs, B., Schmid, C.: Human detection using oriented histograms of flow and appearance. In: Leonardis, A., Bischof, H., Pinz, A. (eds.) ECCV 2006. LNCS, vol. 3952, pp. 428–441. Springer, Heidelberg (2006)

47. Wojek, C., Walk, S., Schiele, B.: Multi-cue onboard pedestrian detection. In: IEEE Conference on Computer Vision and Pattern Recognition (2009)
48. Ojala, T., Pietikainen, M., Maenpaa, T.: Multiresolution grayscale and rotation invariant texture classification with local binary patterns. IEEE Trans. Pattern Anal. Mach. Intell. **24** (7), 971–987 (2002)
49. Rodriguez, Y.: Face detection and verification using local binary patterns. Ph.D. thesis, EPF Lausanne (2006)
50. Zheng, Y., Shen, C., Hartley, R.I., Huang, X.: Effective pedestrian detection using center-symmetric local binary/trinary patterns. In: CoRR (2010)
51. Wang, X., Han, T.X., Yan, S.: An HOG-LBP human detector with partial occlusion handling. In: IEEE International Conference on Computer Vision (2009)
52. Ott, P., Everingham, M.: Implicit color segmentation features for pedestrian and object detection. In: IEEE International Conference on Computer Vision (2009)
53. Hussain, S., Triggs, B.: Feature sets and dimensionality reduction for visual object detection. In: British Machine Vision Conference (2010)
54. Schwartz, W., Kembhavi, A., Harwood, D., Davis, L.: Human detection using partial least squares analysis. In: IEEE International Conference on Computer Vision (2009)
55. Wojek, C., Schiele, B.: A performance evaluation of single and multi-feature people detection. In: Rigoll, G. (ed.) DAGM 2008. LNCS, vol. 5096, pp. 82–91. Springer, Heidelberg (2008)
56. Walk, S., Majer, N., Schindler, K., Schiele, B.: New features and insights for pedestrian detection. In: IEEE Conference on Computer Vision and Pattern Recognition (2010)

Wearable ECG SoC for Wireless Body Area Networks: Implementation with Fuzzy Decision Making Chip

Manikandan Pandiyan[1(✉)] and Geetha Mani[2]

[1] Mercedes-Benz Research and Development India, Bangalore, India
vanajapandi@gmail.com
[2] School of Electrical Engineering, Vellore Institute of Technology,
Vellore, India
geethamr@gmail.com

Abstract. The work aims to present an ultra-low power Electrocardiogram (ECG) on a chip with an integrated Fuzzy Decision Making (FDM) chip for Wireless Body Sensor Networks (WBSN) applications. The developed device is portable, wearable, long battery life, and small in size. The device comprises two designed chips, ECG System-on-Chip and Fuzzy Decision Maker chip. The ECG on-chip contains an analog front end circuit and a 12-bit SAR ADC for signal conditioning, a QRS detector, and relevant control circuitry and interfaces for processing. The analog ECG front-end circuits precisely measure and digitize the raw ECG signal. The QRS complex with a sampling frequency of 256 Hz is extracted after filtering. The extracted QRS details are sent to the decision maker chip, where abnormalities/anomalies in patient's health are detected and an alert signal is sent to the patient via wireless communication protocol. The patient's ECG data is wirelessly transmitted to a PC, using ZigBee or a mobile phone. The chip is prototyped and employed in a standard 0.35 μm CMOS process. The operating voltage of Static RAM and digital circuits and analog core circuits are 3.3 V and 1 V, respectively. The total area of the device is about 6 cm^2 and consumes about 8.5 μW. Small size and low power consumption show the effectiveness of the proposed design, suitable for wireless wearable ECG monitoring devices.

1 Introduction

According to World Health Organization (WHO), cardiovascular and modern human behavior-related diseases are the major cause of mortality worldwide. These types of cardiovascular related-diseases, like Cardiac arrhythmias, Atrial fibrillation, and Coronary heart diseases, can be monitored and controlled with continuous personal healthcare supervision [1, 2]. Electrocardiogram (ECG) embodies the cardiovascular condition, therefore, is considered one of the most important human physiological signals. In applying measurement of physiological signals for continuous monitoring, patients usually cannot carry a bulky instrument, which restricts their mobility and makes them uncomfortable, with so many electrodes and cables attached to their bodies. Therefore, there is growing demand for a compact wearable ECG acquisition

© IFIP International Federation for Information Processing 2016
Published by Springer International Publishing AG 2016. All Rights Reserved
Y. Shin et al. (Eds.): VLSI-SoC 2015, IFIP AICT 483, pp. 67–86, 2016.
DOI: 10.1007/978-3-319-46097-0_4

system [2]. Wearable monitoring devices can record physiological variables, like ECG, blood pressure, etc. for several hours and store them in the memory for future use. The stored ECG data can then be utilized by clinicians or cardiologists for further diagnosis.

The graphical embodiment of a wearable system for continuous remote monitoring is illustrated in Fig. 1. Wearable sensors/electrodes (deployment in accordance with the clinical application) collect the physiological signals for monitoring the patient's health status. These wearable sensors continuously monitor vital signs, like heart rate and blood pressure, when the patient with chronic heart disease is undergoing clinical involvement. Wearable devices are also applied in home-based rehabilitation interventions for continuous personal health monitoring. Wireless protocols can be integrated with wearable systems to facilitate long-term health monitoring for patients diagnosed with cardiac diseases. The wireless communication is relied upon and used to transmit the physiological data continuously to a central place (an access point or a mobile) and to remote central (server or emergency centre) via internet. In emergency situations, an alarm/alert signal can be transmitted to the remote emergency centre for facilitating medical assistance to patients. Family members or clinicians are also alarmed when the patient is in an emergency condition through the technology and enabled to monitor the patient's medical status continuously. Even though there are advantages of wearable devices, many future challenges should be addressed. This primarily requires the support of innovative sensor technologies, especially Wireless Body Sensor Networks (WBSN), formed with various wearable biomedical sensors.

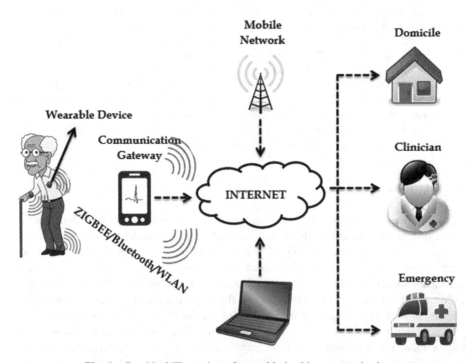

Fig. 1. Graphical illustration of wearable health care monitoring

Since the constraints on battery life and form factor are crucial, these sensors have a very stringent power requirement. To aid low cost, ultra-low power design is essential for developing wearable devices. In terms of cost, size, and performance, System-on-Chip (SoC) implementation is an attractive option.

In this chapter, the development and deployment of the wearable ECG SoC monitoring system are studied, regarding key technology perspective. The following sections present the prior art and essential components of wearable devices, System overview, Proposed ECG SoC, and Fuzzy Decision Maker Chip. Concluding remarks, observations, and future reservations are discussed in the final section.

2 Prior Art

Wireless Body Area Network (WBAN) is the fundamental component of a wireless ECG monitoring system. WBAN allows the integration of various other components, like intelligent systems, miniaturized components, low-power sensor nodes, etc. Therefore, the combination of SoC concepts, wearable technology, Wireless Sensor Network (WSN), and research in artificial intelligence produce novel approaches, resulting in better health care services. System-on-a-chip (SoC) is a felicitous option for device development because of its small size, low power consumption, and lower cost features. Developing SoC for Wireless Body Area Network applications intends to carry healthcare monitoring closer from clinical intervention to domiciles. It allows physiological signal monitoring to be conducted more regularly than limiting it to hospitals or clinics. WBSN is foreseen as the next generation health care monitoring platform, as it is considered a reliable, low-cost high-patient-safety health care monitoring system. In recent years, the development of ECG SoC for WSBN applications has attracted much attention [2, 14]. A wearable monitoring system is proposed in [3] to monitor various physiological variables, such as ECG, blood pressure, and temperature. Also, the Global Positioning System (GPS) co-ordinates of patient or wearer with the acquired variables are transmitted wirelessly to a remote station.

Targeting patients with chronic high-risk heart/respiratory diseases, a wrist worn wearable medical monitoring and alert system (AMON) monitors physiological variables. For terrestrial and space applications, physiological parameters of the astronauts in space should be monitored continuously. To address the mentioned problem, a wearable system, called 'Life Guard', is proposed [4] to monitor the health status of astronauts. The deployment of a biopatch with integrated low-power SoC prototype is proposed by Yang et al. [5] to facilitate features, such as a three-stage front-end signal conditioning circuit, 8-bit successive-approximation-register (SAR) ADC, and a digital core. An integrated wireless ECG SoC for WBSN applications is proposed in [6], which comprises a two-channel ECG front-end, an 8-bit SAR-ADC, a simple microcontroller, a SRAM memory, and RF-transceivers. Many ECG SoCs implementations for WBSN applications employ a microcontroller or microprocessor to establish the remote gateway [7, 14]. In worst cases, there is a need for an artificial intelligence approach, integrated with wearable ECG SoC, when abnormal ECG episodes are to be detected instantly. This solution addresses diseases, like cardiac arrhythmia or silent myocardial ischemia, to be easily identified for clinical treatment. This increases the need for low cost and easy to use wearable wireless ECG sensors with integrated decision making to alert personnel. The following sections narrate about an ultra-low power ECG on Chip with integrated CMOS Fuzzy Decision Making Chip that addresses the issues in existing solutions.

3 Wearable ECG System: With Decision Making

3.1 System Overview

The proposed health care architecture includes two parts: (a) Main unit and (b) Remote unit, as shown in Fig. 2. The Main unit contains wearable textile electrodes, designed ECG front-end chip, FDM chip, a controller, and a ZigBee transceiver. The remote unit (personal gateway) can be a mobile phone or a personal computer with an USB interface. The main unit records the ECG from wearable textile electrodes and wirelessly transfers the data to a remote unit. The designed ECG acquisition chip for low power use is described in the next sub-section. The ECG acquisition chip comprises: (1) specially designed textile electrodes for acquiring the ECG; (2) a miniature printed circuit board with ECG front end circuits; (3) Analog to Digital Conversion unit; (4) QRS Detection; and (5) System control unit. The ECG data is buffered, using low power microcontroller internal memory to minimize power consumption before wirelessly sending it to the remote unit. The main unit also performs the other tasks, such as system initialization, data buffering, and scheduling wireless communication. The Fuzzy Decision Making (FDM) chip (3 × 3 fuzzy controller; nine rules are accessible) takes decisions when necessary. Depending on applications, control voltages set on IC pins change the rules of fuzzy inference. The study of the Fuzzy chip is explained in detail in subsequent sections. ZigBee protocol is chosen as a wireless communication protocol (TI CC2420) to provide reasonable power consumption and adequate data rate. The prototype uses a low power TI MSP430 microcontroller for data management, wireless ZigBee baseband, and routing management. The prototype model is designed for patients, regarding comfort and ease of use, thus, not affecting regular activities of patients. In addition, the entire unit is sealed within a smart textile shirt. So, the patient can wear and remove it easily.

For various medical applications, the acquired physiological variables should be analyzed continuously. The remote unit (personal gateway) can be a mobile phone or a personal computer with an USB interface. The important functions of the remote unit are receiving the data wirelessly, database management, ECG analysis, graphical user interface (GUI) interface, and customization. To avoid signal interference from other wireless devices, the remote unit has specific authentication to process the received data. In addition, there are several options in the GUI interface for customization.

3.2 ECG on Chip

A. ECG Analog Front-End Amplifier. The ECG front-end amplifier is mainly responsible for noise suppression, signal conditioning, and amplification, which

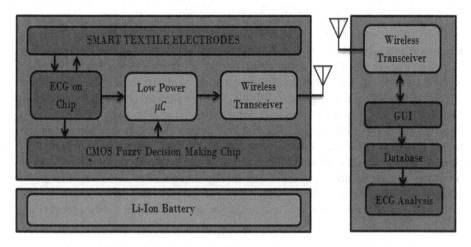

Fig. 2. Block diagram of proposed healthcare architecture

comprises two phases as shown in Fig. 3, namely, low noise AFG with band pass function and a programmable gain amplifier (PGA) to amplify the acquired ECG signals (from textile electrodes), with amplitude in a few millivolts, adopting a flip-over-capacitor technique. The Low noise amplifier not only acts as a preamplifier, but also acts as a band pass filter function with bandwidth between 0.3 and 100 Hz. In the AFG design, two switches (S1 and S2) are integrated to settle down quicker when power is applied, due to the large resistance by the pseudo-resistors. The speeding up of the AFG is done by a reset signal with an appropriate switch during startup of the system.

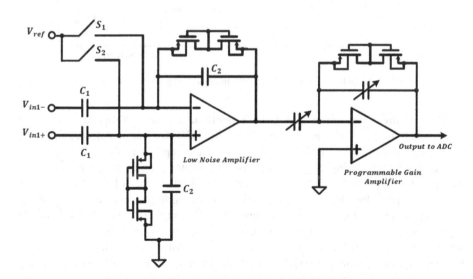

Fig. 3. Typical circuit diagram of ECG front-end low noise amplifier

B. Analog to Digital Converter (ADC). Successive Approximation Register (SAR) ADC is chosen for this WBSN application because of its moderate accuracy and low power overhead. Figure 4 depicts the architecture of the SAR ADC, adopted from literature. The analog ECG output is driven directly by the preceding buffer stage, without the need of an additional hold amplifier, sampled through a bootstrapped switch and held in the capacitive 12 bit DAC, and is then used by open-loop Sample/Hold. The reason for open-loop Sample/Hold is to obtain low power, low cost, fast settling, and less offset error. The obtained analog ECG data are being compared with a reference (REF) and then level-shifted by the DAC. The fixed reference REF helps to compensate the dynamic offset error at the comparator. An on-chip crystal oscillator is used to drive the logic and timing sequence for achieving low power consumption and low jitter. The resultant digital codes are passed to a System Control Unit (SCU) after level conversion and to QRS complex detector for data processing.

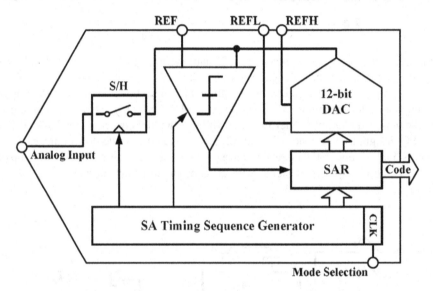

Fig. 4. Schematic architecture of successive approximation ADC

C. Heart Rate Calculation and QRS Detection. The morphological filter [8] is adopted to reduce the noise artifacts present in the ECG data and to detect\estimate the QRS complex details and R-R intervals. The filter comprises a pair of Opening and Closing operations, using dilation and erosion operators, which suppress peaks and valleys. The flowchart for QRS complex detection is illustrated in Fig. 5. The impacts of wandering baseline drift are eradicated by subtracting the mean result of operations (opening and closing) with the original input. The ECG samples are loaded serially into the shift register and then added/subtracted (for dilation/erosion respectively) with the structure element $g(x)$ [8]. The results are compared continuously, using a comparator tree to find the minimum/maximum (for dilation/erosion respectively). A moving average filter (serial structured) is used to reduce the impulse noises and smooth the filtered signal. The received signals are continuously compared against the adaptive threshold and monitored to detect the R-peaks. The current threshold is updated regularly when a new R-peak is identified. By counting the number of clocks between R peaks using a binary counter, R-R interval is measured. The Heart Rate (HR) variable is also calculated by simply counting the number of R-peaks in the last sixty seconds. A parallel-to-serial converter is integrated with the wearable system for transmitting the HR variable through the SPI interface.

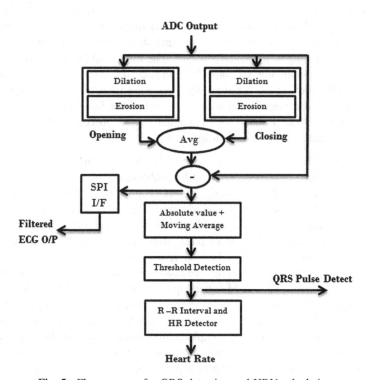

Fig. 5. Flow process for QRS detection and HRV calculation

D. System Control Unit and SPI Interface. System Control Unit (SCU) is solely responsible for generating the interface control signals, based on the host or main controller commands for all the blocks in ECG on-chip. Data framing, CPU interrupt handling, etc. are done by the SCU system, based on control signals, which are generated by state machine. In-order to interface the chip with various host CPUs, the System Control Unit uses an asynchronous FIFO with 8 Kb buffers. Data from the ADC and QRS block is continuously written into the FIFO at the sampling frequency of 256 Hz. Based on the FIFO status, FIFO write/read controllers generate many status signals, which are "full", "nearly full", "empty", and "nearly empty". A microcontroller is employed externally to communicate with the proposed wearable device via a duplex SPI communication interface. The data link transmits QRS complex codes and ECG details with the internal FIFO status flags.

The control vector from external microcontroller is delivered to the internal control registers by the command link.

3.3 Fuzzy Decision Making Chip: Concepts, Design and Implementation

In real world worst-case scenarios, there is a need for an artificial intelligence approach, integrated with a wearable ECG SoC, for detecting abnormal ECG episodes instantly. Especially, diseases like cardiac arrhythmia or silent myocardial ischemia should be identified immediately to alert the clinicians/family members for assistance. Therefore, a Fuzzy Classifier chip meets the critical requirements of medical applications: no delay in response, reliable, high-safety, and low cost. The functional blocks of the FDM chip are detailed in the following sections. It comprises three parts: fuzzifier, inference engine, and defuzzifier. In the fuzzifier, input variables (non-fuzzy) are mapped to the input membership functions. The inference engine handles fuzzy inference, depending on the inference method. Finally, the defuzzifier is used to convert the fuzzy output values from inference engine to non-fuzzy values.

Fuzzy Interface. The implemented architecture is a two-input one-output fuzzy supervisor. Each input has three trapezoidal membership functions or linguistic terms abbreviated as L (Low), M (Medium), and H (High), while the output variable is characterized by singletons. The parameters used for determining membership functions (V_{r+}, V_{r-}, V_{c+}, V_{c-}) are calibrated by voltages applied on FDM IC pins. Since input membership functions have three parts, architecture is a 3×3 fuzzy controller, and nine rules are accessible. Depending on various applications, Control voltages set on IC pins can change these rules. The controller architecture in Fig. 6 is constructed with CMOS components, such as Membership function generator (MFG), MIN circuits and a defuzzifier (D blocks) circuit. A ramp function generation circuit is used for MFG. In fuzzy interface, three basic circuits are used: a ramp generator (RG) circuit [8], a minimum circuit, and a fuzzy complementary circuit. The membership functions for input and output variables of a controller are built, using two ramp functions. In Fig. 7, a form of trapezoidal membership function, using a ramp function and its parameters, is depicted.

The membership functions are generated for input and output variables of a fuzzy decision supervisor. It shows that two ramp functions are necessary to build a membership function. The triangular membership function is a special case of trapezoidal membership function, when c and d coincide. The slope of the ramp functions and position of the membership functions must be tunable. Changing the parameters, a, b, c, and d allow the construction of different membership functions.

Figure 8a shows the ramp generator circuit for membership function generation. It is assumed that all fuzzy sets are normalized. $\text{Supremum}_x (x) = 1$. In RG circuit, the output current i_0 is a function of v_1 and v_2. Considering I_{ss} fixed, and M3, M4, M5, and M6 are matched,

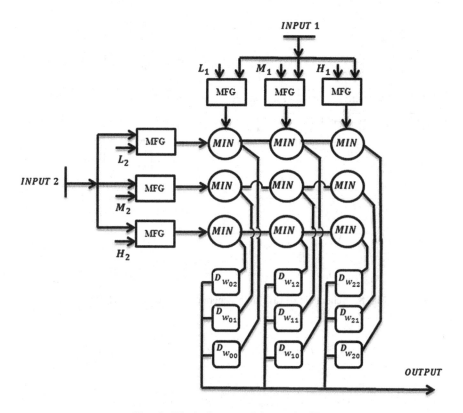

Fig. 6. Block diagram of fuzzy classifier

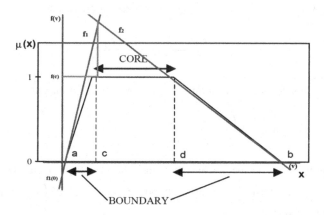

Fig. 7. Trapezoidal membership function and its parameters

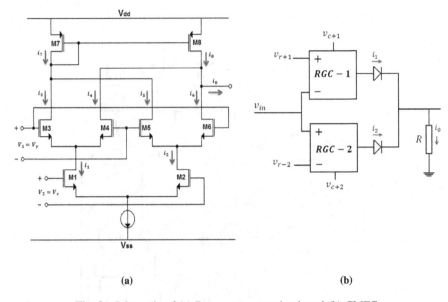

(a)   (b)

Fig. 8. Schematic of (a) Ramp generator circuit and (b) CMFG

$$i_0 = i_3 + i_5 - (i_4 + i_6) = g_{m3}\frac{v_r}{2} - g_{m5}\frac{v_r}{2}\left(-g_{m4}\frac{v_r}{2} + g_{m6}\frac{v_r}{2}\right)$$

$$= \frac{v_r}{2}\left(\frac{I_1}{v_{gso3} - v_t} - \frac{I_2}{v_{gso5} - v_t} + \frac{I_1}{v_{gso4} - v_t} - \frac{I_2}{v_{gso5} - v_t}\right)$$

$$= \frac{v_r}{2}\left(\frac{2I_1}{v_{gso3} - v_t} - \frac{2I_2}{v_{gso5} - v_t}\right),$$

where, all $v'_{gso}s$ are gate to source voltages in quiescent point, and also

$$I_1 = \frac{I_{ss}\frac{v_c}{2}}{v_{gso1} - v_t}, \quad I_2 = \frac{-I_{ss}\frac{v_c}{2}}{v_{gso2} - v_t}$$

Using previous equations, assuming M1 and M2 are also matched, then

$$i_0 = \frac{I_{ss}}{v_{gso1} - v_t}\left(\frac{1}{v_{gso3} - v_t} + \frac{1}{v_{gso5} - v_t}\right)v_r v_c$$

Assuming v_c is set to be a constant value, hence $v_{gso1} - v_t$ is constant. Also, changes of $v_{gso3} - v_t$ and $v_{gso5} - v_t$ are in opposite directions. So $\left(\frac{1}{v_{gso3} - v_t} + \frac{1}{v_{gso5} - v_t}\right)$ is approximately constant.

Therefore, $i_0 = kv_1v_2$
where,

$$k = \frac{I_{ss}}{v_{gso1} - v_t}\left(\frac{1}{v_{gso3} - v_t} + \frac{1}{v_{gso5} - v_t}\right)$$

The RG circuit can generate both a positive and negative slope ramp function. Suppose $v_{c-} = 0, v_{c+}$ and v_{r-} are constant, then

$$i_0 = k(v_{r+} - v_{r-})(v_+) = kv_{c+}v_{r-} - kv_{c+}v_{r-} = mv_{r+} - n$$

where,

$$m = kv_{c+}, n = kv_{c+}v_{r-}$$

Thus, a positive slope ramp is generated.
Similarly, if $v_{c+} = 0$, and let v_{c-}, v_{r-} be constant, then

$$i_0 = k(v_{r+} - v_{r-})(-v_{c-}) = -kv_{c-}v_{r+} + kv_{c-}v_{r-} = -mv_{r+} - n$$

where,

$$m = kv_{c-}, n = kv_{c-}v_{r-}$$

Therefore, by changing v_{r+}, v_{r-} and v_c, different ramp functions are generated. It is evident that the output voltage of RG circuit

$$v_0 = R_{out}i_{out}$$

where R_{out} is output resistance.

Two ramp functions f_1 and f_2 with positive and negative slopes, respectively, are shown in Fig. 7. Assuming RG circuit has generated these functions, f_1 and f_2 can be equated as below

$$f_1 = m_1v - n_1$$

$$f_2 = n_2 - m_2v$$

To construct a tunable membership function,

$$f_1(0) = -n_1 = -kv_{r1-}v_{c1+}$$

$$f_2(0) = n_1 = kv_{r2-}v_{c2-}$$

$$a = v|_{f_1=0} = \frac{n_1}{m_1} = \frac{kv_{r1-}v_{c1+}}{kv_{c1+}} = v_{r1-},$$

$$b = v|_{f_2=0} = \frac{n_2}{m_2} = \frac{kv_{r2-}v_{c2+}}{kv_{c2+}} = v_{r2-},$$

When $f_1 = f_2$, can be written as $m_1c - n_1 = -n_2 + m_2c$
Therefore,

$$c = \frac{n_1 - n_2}{m_1 - m_2} = \frac{v_{r1-}v_{c1+} - v_{r2-}v_{c2-}}{v_{c1+} - v_{c2-}}$$

With this value for c,

$$f(c) = f_1(c)$$

$$= f_2(c) = \frac{kv_{c1+}v_{c2-}}{v_{c1+} - v_{c2-}}(v_{r1-} - v_{r2-})$$

To construct a fuzzifier with desirable capabilities, Set $v_{r1-} = v_{r2-}$. A reverse triangular in the positive region of vertical axis will be formed as shown in Fig. 7. By clipping this triangular area with a constant value E, a controllable membership function can be obtained.

In this case,

$$a = v|_{f_2=E} = \frac{n_2 - E}{m_2} = v_{r2-} - \frac{E}{kv_{c2-}},$$

$$b = v|_{f_1=E} = \frac{E - n_1}{m_1} = \frac{E}{kv_{c1+}} - v_{r1-},$$

Varying v_{c1} and v_{c2} results in a change in a and b. For $v_{r1} = v_{r2}$, it will also be variable. The position of the membership function can be changed. Note that a trapezoidal membership function is obtained when $v_{c+} > v_{c-}$ and in this case, $c = v_{c-}$ and $d = v_{c+}$. Considering these points, it is necessary that c and d are selected equal to v_{r1} and v_{r2}, respectively, and then by changing v_{c1} and v_{c2}, a and b are determined. Diodes are used to eliminate the extra parts of membership function. Since two ramp functions are needed, two RG circuits are used for each membership function. The typical circuit diagram of complementary membership function generator (CMFG) is shown in Fig. 8b and adapted from [8]. Considering the output voltage $v_{0ut} = Ri_o$ where,

$$i_0 = \begin{cases} i_1 = k(v_{1+} - v_{in})v_{c1+}; v_{r1} + fv_{in} \\ i_2 = k(v_{in} - v_{r2-})v_{c2-}; v_{r2}-pv_{in} \end{cases}$$

Minimum circuits can be used to clip the extra curves and then reverse it by a fuzzy complementary circuit (FCC). Figure 10 illustrates the circuit diagram of a fuzzy complementary circuit. In FCC, the two inputs are connected to the gates of the transistors. A multi-input minimum circuit with over two transistors is needed in

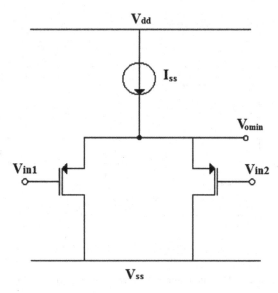

Fig. 9. Schematic representation of minimum circuit

applications, such as three-input fuzzy controllers. As illustrated in Fig. 9, the output voltage $v_{o(min)}$ always takes the smaller value of two inputs v_{in1} and v_{in2} with a positive offset voltage v_{offset}.

$$v_{o(min)} = \min(v_{in1}, v_{in2}) + v_{offset}$$

This result $v_{o(min)}$ specifies that the offset voltage is about 1 V. To compensate the offset voltage, v_{off} a negative level shifter circuit is necessary. The Negative shift is achieved by a fuzzy complement circuit with a reference voltage of E. The complementary membership function must be converted to an ordinary membership function, achieved by a fuzzy complementary circuit of two g_m circuits.

From Fig. 10, assuming the reference voltage E = 0,

$$v_{comp} = R_i = R(i_1 + i_2) = R(-g_m v_{in} - g_m v_{comp})$$

Therefore, this is opposite to the input voltage. The complement of the membership function is $E \neq 0$,

$$v_{comp} = E - v_{in} \quad if \ Rg_m \gg 1$$

Note that, in the final fuzzifier structure, E must be set to compensate offset voltage, associated with the minimum circuit. The attenuation is due to the limited gain of Rg_m that is equal to $-Rg_m = 1 + Rg_m$. This attenuation is the same in each one of the fuzzifiers used in the controller, and the error due to attenuation does not affect fuzzy processing considerably. If the input signal has a negative DC value, then E must be chosen to be greater than the Supremum value by $|V_{dc}|$.

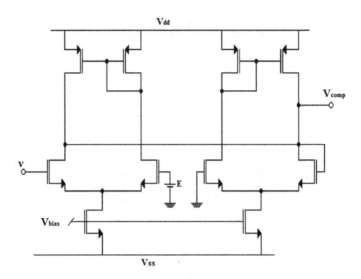

Fig. 10. Schematic diagram of fuzzy complementary circuit

Inference Engine. With Mamdani's inference technique, the inference engine is accomplished by a set of intersection and union operations [10, 12]. The Min- Product inference method is chosen for inferencing in which a minimum of two inputs can be specified. Nine two-input minimum circuits in the inference engine are needed. Therefore, the controller has two inputs with three membership functions. The Min blocks [10] and their synthesis in the complete controller structure will do the inferencing.

Defuzzification. A novel defuzzifier is used [8, 10] in which the center of the area is calculated without employing a division circuit. Therefore, it occupies a small chip area. The main idea is based on parallel conductances g_n, stating implicitly that the output voltage of the defuzzifier circuit is the average value of the inputs. The contribution of each input to the output is weighted by the conductance g_i that is a controllable variable. The below equation provides a non-fuzzy or defuzzifier output.

$$V_{defuzz} = \frac{g_1 V_1 + g_2 V_2 + \ldots + g_n V_n}{g_1 + g_2 \ldots \ldots + g_n}$$

MOS transistor is used as a controllable g-element in this work. Offset voltage v_t in a MOS transistor must be cancelled to control g_{ds} by v_{gs} $(g_{ds} = k v_{gs})$ linearly.

$$g_{ds} = k\left(v_{gs} - v_t\right); \; k = \frac{1}{2}\mu c_{ox}\frac{W}{L}$$

Level shifter circuit (LSC) is used to compensate offset voltage [8]. The complete schematic diagram of defuzzifier is depicted in Fig. 11.

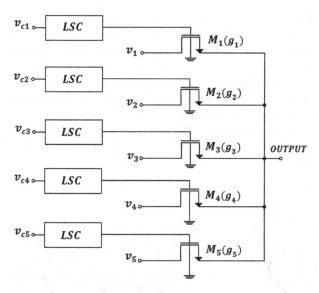

Fig. 11. Schematic illustration of defuzzifier

4 Results and Discussion

The wearable ECG sensor node system fits perfectly on a shirt. The main unit provides a versatile framework for incorporating sensing, monitoring, and information processing devices. The designed wearable device can be deployed in a variety of applications, such as public safety, health monitoring, and sports. The vital signal monitoring functionality of the smart shirt is tested in real time. The inference performance test is done, based on physical activity under various conditions. The abnormal ECG signal is measured and stored in the fuzzy inference engine. The fuzzy decision making rules are framed in such a way that, when an abnormal ECG signal is detected, an alert signal is sent promptly to the remote gateway via microcontroller. A wearable smart shirt transfers the physiological ECG signals over a wireless sensor network at the test. The following sections depict the important results of the proposed system.

4.1 ECG Acquisition

To ensure comfort, the clothing is designed from a knitted conductive textile fabric for reducing flammability. Rectangles of electrically conductive textile fabric in knitted design were stitched on the position of the pectoral muscles [13]. The conductive textile fabric is realized from a blended yarn of the composite containing silver nanoparticles, which provide electrical conductivity of the yarn and the resultant knitted fabric. The content of silver nanoparticles provides corrosion resistance of textile electrodes, antibacterial and anti-allergic properties, and mechanical and electrical stability when exposed to sweat. The design of blended conductive textile fabric, made from conductive yarn, enables traditional maintenance of T-shirts (washing, ironing) and long-term stability of surface conductivity of the electrodes with a high number of wash cycles. The designed conductive textile fabrics are circular in shape, with dimensions 5 × 5 cm. Figure 12 shows the wearable electrodes, which comprise a conductive fabric electrode pair and the wearable sensor node system placed on the wearer's chest placement. To provide a sufficient potential difference, the electrodes are positioned 100 mm apart. The ECG measurements are obtained through wearable textile electrodes (Zero Resistance, 100 % Silver fiber for conductive part), and the acquired ECG signals are processed, using other modules in this health architecture for decision making.

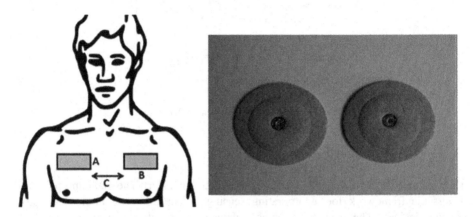

Fig. 12. Placement of wearable textile electrodes

4.2 Fuzzy Decision Making

In the FDM module, the Heart rate variability (HRV) is computed from the time series R-R intervals (R-peak to R-peak), converted into a uniformly sampled time-spaced sequence. As the physiological condition (wakefulness state, sleep state, etc.) of a patient changes, the power spectral density (PSD) of heart rate differs. The low-to-high (L/H) frequency ratio is considered an effective assessment and indicator of such change, because it reflects the balancing action of the sympathetic and parasympathetic nervous-system branches. Power Spectral Density of HR variations is calculated, and the three frequency bands, such as Very low frequencies (VLF: 0–0.04 Hz), Low frequencies (LF: 0.04–0.15 Hz), and High frequencies (HF: 0.15–0.5 Hz), have been utilized. The features extracted from HRV and PSD are used to feed the fuzzy logic engine that computes epoch-by-epoch (30 or 60 s per period) inferences. The fuzzy inference rules are based on the observed details of normal and abnormal ECG signals. The inputs of FDM chip are details of QRS complex and PSD results. The ranges of membership functions are tunable by changing the voltages of FDM IC pins, which is done by the developer via microcontroller.

The FDM chip provides the index values (i.e., defuzzified output in the range of 0–2), which is sent to the microcontroller. Hence, there are output states, such as normal, sleep onset/fatigue, and abnormal, decoded by microcontroller unit. A set of meaningful rules has been framed. The following are the strongest:

IF HR Variability is Low **AND** LF/HF ratio is Medium
THEN the Output signal is SLEEP_ ONSET
IF HR Variability is High **AND** LF/HF ratio is High
THEN the Output signal is NOR_WAKE
IF HR Variability is Low **AND** LF/HF ratio is Low
THEN the Output signal is DROWSY

Figure 13 shows the major difference between normal and abnormal ECG signals, noting that an abnormal ECG data has an elevated T wave.

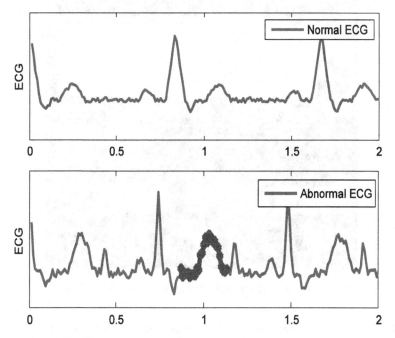

Fig. 13. Measured sample ECG data for inference engine

4.3 Performance Evaluation

The status is continuously sent to the remote unit every 2 min or preset time in the controller. The results of the experiments, as shown in Table 1, confirmed our hypothesis that human health status can be predicted by the FDM module through extracted ECG features. Detection succeeds, based on ECG signal captured from the wearable textile electrodes. When the signal is sensed, the system detects the status, and if abnormal, an alert signal is transmitted. Measurement accuracy of Fuzzy based ECG classification confirms to be robust enough to perform over 95 % successful early detections. Therefore, the proposed system can make decisions, based on the acquired ECG data.

Figure 14 depicts the designed graphical user interface for the proposed architecture. Timing of the early detection capability of each system is also evaluated during the tests.

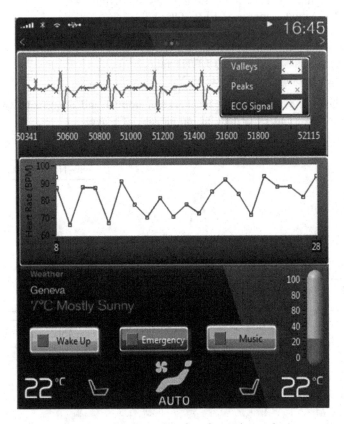

Fig. 14. Designed graphical user interface for testing and measurements

Table 1. Performance evaluation of fuzzy decision making in the proposed system

Rule-check	Clinical datasets			
	Number of data sets used for testing	Number of data sets correctly classified	Number of data sets wrongly classified	Accuracy (%)
Drowsiness	21	19	2	90
Sleep onset	19	18	1	94
Normal	23	23	0	100
Rule-check	Simulated datasets			
	Number of data sets used for testing	Number of data sets correctly classified	Number of data sets wrongly classified	Accuracy (%)
Drowsiness	89	82	7	92
Sleep onset	102	98	4	96
Normal	213	213	0	100

5   Concluding Remarks

A wireless ECG on a chip with an integrated Fuzzy Decision making system is proposed for real-time ECG health monitoring. The proposed wearable device is small, user-friendly, has a long battery life, and is capable of wirelessly transmitting ECG data continuously to a remote station for detailed diagnosis. The FDM chip is integrated with ECG on Chip to take the decisions for alerting the patients when necessary. The designed FDM responds immediately when anomalies are found in ECG data. The proposed device has already been tested with a reference high-quality measurement system for verification of accuracy and showed that the accuracy of the proposed device is good enough, and the variation in key ECG parameters obtained from the proposed device and the reference device is acceptable for clinical usage.

Acknowledgement. This research was financially supported by International Society of Automation (ISA) Educational Foundation Scholarship 2014. We would like to show our heartfelt gratitude to editors and reviewers for sharing their pearls of wisdom with us during manuscript preparation. We also thank Natarajan Sivaraman, Asst. Prof., Dept. of ICE, PSG Tech for the valuable comments that greatly improved the manuscript.

References

1. Chen, C.Y., Chang, C.L., Chang, C.W., Lai, S.C.T., Chien, F., Huang, H.Y., Chiou, J.C., Luo, C.H.: A low-power bio-potential acquisition system with flexible PDMS dry electrodes for portable ubiquitous healthcare applications. Sensors **13**, 3077–3091 (2013)
2. Pandiyan, M., Mani, G., Jerome, J., Sivaraman, N.: Integrating wearable low power CMOS ECG acquisition SoC with decision making system for WSBN applications. In: 2015 IFIP/IEEE International Conference on Very Large Scale Integration (VLSI-SoC), pp. 154–158, 5–7 October 2015
3. Pandian, P.S., et al.: Wearable multi-parameter remote physiological monitoring system. Med. Eng. Phys. **30**(4), 466–477 (2007)
4. Anliker, U., Ward, J.A., Lukowicz, P., et al.: AMON: a wearable multi parameter medical monitoring and alert system. IEEE Trans. Inf. Technol. Biomed. **8**(4), 1–11 (2004)
5. Yang, G., Xie, L., Mantysalo, M., Chen, J., Tenhunen, H., Zheng, L.R.: Bio-patch design and implementation based on a low-power system-on-chip and paper-based inkjet printing technology. IEEE Trans. Inf. Technol. Biomed. **16**(6), 1043–1050 (2012)
6. Khayatzadeh, M., Zhang, X., Tan, J., Liew, W.S., Lian, Y.: A 0.7-V 17.4-ψW3-lead wireless ECG SoC. IEEE Trans. Biomed. Circ. Syst. **7**(5), 583–592 (2013)
7. Halin, N., Junnila, M., Loula, P., Aarnio, P.: The life shirt system for wireless patient monitoring in the operating room. J. Telemed. Telecare **11**, 41–43 (2005)
8. Peyravi, H., Khoei, A., Hadidi, K.: Design of an analog CMOS fuzzy logic controller chip. Fuzzy Sets Syst. **132**, 245–260 (2002)
9. Lobodzinski, S.S., Laks, M.M.: New devices for very long-term ECG monitoring. Cardiol. J. **19**(2), 210–214 (2012)
10. Pandiyan, M., Mani, G.: Embedded low power analog CMOS fuzzy logic controller chip for industrial applications. In: 2015 IFIP/IEEE International Conference on Very Large Scale Integration (VLSI-SoC), pp. 43–48, 5–7 October 2015

11. Van Helleputte, N., Tomasik, J.M., Galjan, W., Mora-Sanchez, A., Schroeder, D., Krautschneider, W.H., Puers, R.: A flexible system-on-chip (SoC) for biomedical signal acquisition and processing. Sens. Actuators A **142**(1), 361–368 (2008)
12. Manikandan, P., Geetha, M., Jerome, J.: Weighted fuzzy fault tolerant model predictive control. In: IEEE International Conference on Fuzzy Systems (FUZZ-IEEE), pp. 83–90. IEEE (2014)
13. Vojtch, L., Bortel, R., Neruda, M., Kozak, M.: Wearable textile electrodes for ECG measurement. Adv. Electr. Electron. Eng. **11**(5), 410–414 (2013)
14. Lopez, G., Custodio, V., Moreno, J.I., Moreno, J.I.: LOBIN: E-textile and wireless-sensor-network-based platform for healthcare monitoring in future hospital environments. IEEE Trans. Inf. Technol. Biomed. **14**(6), 1446–1458 (2010)

Delay Testing Based on Multiple Faulty Behaviors

Masahiro Fujita$^{(\boxtimes)}$

VLSI Design and Education Center, The University of Tokyo, 2-11-16 Yayoi,
Bunkyo-ku, Tokyo, Japan
fujita@ee.t.u-tokyo.ac.jp

Abstract. We discuss overall "observed" behaviors of circuits due to
additional delays caused by various variations in the chips and propose
delay testing methods based on such analysis. First we examine functional changes caused by the additional delays on the inputs of each gate
in the circuit. We show that unlike structural faults, e.g., stuck-at faults,
such additional delays can introduce many more different faulty functions on a gate, and we propose two functional delay fault models for the
changed behaviors caused by the additional delays, one with one time
frame and the other with two time frames. As such additional delays
by variations and other reasons naturally happen in multiple locations
simultaneously, there can be exponentially many multiple fault combinations to be considered. It is not at all easy to analyze them with traditional automatic test pattern generation (ATPG) methods which rely
on fault dropping with explicit representation of fault lists. So in the
second part of the paper, we present an ATPG method based on implicit
representations of fault lists. As faults are represented implicitly, even
if numbers of simultaneous faults are large and total numbers of fault
combinations are exponentially many, we may still be able to successfully
perform ATPG processes. Experimental results have shown that even for
large circuits in the ISCAS89 benchmark circuits, complete sets of test
vectors for all multiple combinations of the proposed functional delay
faults are successfully generated in a couple of hours. The numbers of
required test vectors for complete testing are surprisingly small, e.g., only
a few thousands for circuits having more than ten thousands of gates,
even though there are more than $2^{(ten\ thousands)}$ combinations of multiple faults in those circuits. This indicates that the proposed multiple
functional delay fault models may have practical values as they consider
all types of multiple functional faults caused by extended delays in the
circuit.

1 Introduction

As the semiconductor technology continues to shrink, we have to expect more
and more varieties of variations in the process of manufacturing in particular for
large chips. Such variations, especially ones on delays in circuits, can change the

© IFIP International Federation for Information Processing 2016
Published by Springer International Publishing AG 2016. All Rights Reserved
Y. Shin et al. (Eds.): VLSI-SoC 2015, IFIP AICT 483, pp. 87–108, 2016.
DOI: 10.1007/978-3-319-46097-0_5

circuits' "observed" functional behaviors, which is generally called delay fault. Here, we discuss such changed functionality caused by the additional delays due to variation and other reasons. Delay testing is getting a lot of attention as there are more and more additional delays possibly happening within a chip in distributed ways, such as accumulated effects of small delays. There have been works on testing whether such delays causes any changes in behaviors of the circuit, which is generally called delay testing. Most of them try to measure delays of the circuit being tested by checking delays of signal propagation paths using two test vectors as shown in Fig. 1, such as testing longest paths [1], analyzing accumulation of small delays in gates [2], and many others. With the two test vectors, specific signal propagation paths are activated, and their actual delays are measured by physical facilities such as LST testers. It is measured and checked whether some paths exceed the maximum allowed amount of delays or not. In this paper, although we discuss long path delay problems only, short path delay problems can be dealt with in a similar way.

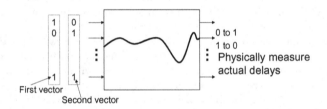

Fig. 1. Measure delays with two test vectors

As there may be so many signal propagation paths in large circuits and delays could vary a lot depending on variations, delay estimation, such as minimum and maximum delays, may have to have large ranges in values. As a result, those delay testing methods may not work well, because appropriate threshold delays for delay testing may not be easily defined, especially when variations in the chips are large and distributed.

In this paper instead of trying to measure or estimate delays, which is a common way in the current delay testing methods, we concentrate on analyzing what are possible functional changes due to such distributed and accumulated delays with wide ranges of values. Our proposed delay fault model is to define the possible situations where inputs of some gates in the circuit could get the values of previous cycles instead of the current cycles due to the increase of delays in the circuit. This fault model is called as FDF2 (Functional Delay Fault with two time frames), as it needs two time frames to define. We also define a simplified functional delay fault model where inputs to gates could get the wrong values rather than the previous values due to widely distributed and additional delays. We call this delay fault model as FDF1 (Functional Delay Fault with one time frame), as it is based on one time frame.

Under FDF2, for a signal in a circuit, if the value in the previous cycle is the same as the one in the current cycle, such additional delay will not introduce any changes in terms of functionality. On the other hand, if they are different, "observed" functionality may change from the original one, or may not change depending on internal don't cares derived from the fanout regions from the faulty locations. If we assume that a gate in a circuit may use the values in the previous cycles as its inputs, the observed and resulting functions realized by the gate can vary in many ways as discussed in the following sections. For example, there are possibly "16" different functions which can be realized by a two-input AND (OR) gate with such additional delays. That is, all possible functions with two-inputs may potentially be observed with a two-input AND (OR) gate with additional delays based on our fault model, FDF2. This may suggest that it may make sense to model faulty behaviors caused by distributed and additional delays as general functional faults rather than structurally defined faults, such as stuck-at faults, although in this paper we use a different and more straightforward way to define the faulty behaviors.

Please note that in the above discussion, for example, an AND gate is assumed to be doing the correct operations all the time, but its input values can become partially or totally wrong due to delays, which is observed as functional changes.

The functional delay fault model, FDF2, is defined over two time frames of sequential circuits, since for an input of each gate in the circuit we need to refer to its previous value as well as its current value, i.e., if it is faulty, use the previous one, and if it is not faulty, use the current one. We also define and evaluate a less accurate but simpler fault model, FDF1. It is a fault model where inputs of a gate get the complemented values of the correct ones under faults. As the values of the same signal may or may not be different in the fault model, FDF2, this simpler model is more conservative in terms of faulty behaviors, i.e., always receiving wrong values in this simplified fault model, if it is faulty.

In both fault models, in order to analyze the delay related faulty behaviors, it is essential to deal with multiple faults rather than single faults. As there is no specific assumption on the variations which cause additional delay, here we assume each input of a gate may have independent accumulated delays from primary inputs and inputs from the flipflops. Such additional delays can happen in multiple locations simultaneously. It may be the case where most of the gates in a circuit may get the values for the previous cycle rather than the current cycle. In such cases, from the viewpoint of faults caused by the delays, there can be many, such as hundred, thousands or more simultaneous faults in the circuit. As a result, when we are generating test vectors for such combinations of faults, we need to manage ultra large lists of fault combinations, since there can be exponentially many fault combinations under multiple fault models.

In general, ATPG (Automatic Test Pattern Generation) processes use fault simulators to eliminate all of the faults combinations which can be detected with the current set of test vectors (called fault dropping process). Traditionally in almost all cases, fault combinations are explicitly represented in fault lists, as that is an easy and simple way for their manipulations. For functional and multiple faults, however, explicit representation is no longer feasible. For example, if there are 16 faults possible with a gate and we need to consider up to 10 simultaneous and multiple faults, the size of the fault list in explicit representations is in the order of 16^{10} or more. This is the case when we consider only one particular set of 10 faulty locations. In general, we need to take care of much more fault combinations.

In general there are many such sets of locations in a circuit. No explicit representation can keep such large numbers of instances. Instead we need to represent them with some sorts of "implicit" methods. This is in some sense a similar problem to so called "state explosion" problem [14] in model checking and formal analysis in general. In those fields, implicit representations are commonly used in order to deal with larger problems. In this paper, we show an ATPG method based on such implicit representation of fault lists based on the techniques first developed in [5]. By formulating the ATPG process as an incremental Satisfiability (SAT) solving, fault lists are naturally represented and processed in implicit ways as logical formulae. We define circuits based on multiplexers in order to represent the two delay fault models, FDF1 and FDF2, discussed above.

Those circuits have parameter variables, and the values of parameter variables determine which faults currently exist or do not exist in the target circuit. Such a circuit for fault modeling is introduced to each possibly faulty location, i.e., all inputs of each gate in the target circuit. Therefore, all the parameter variables altogether show how multiple faults exist in the circuit. This is an implicit way to represent multiple faults. As faults are represented implicitly, even if numbers of simultaneous faults are large, such as $2^{(ten\ thousands)}$, we can still successfully perform ATPG processes as shown in the experiments below.

The rest of the paper is organized as follows. In the next section we discuss possible functional faults or wrong operations caused by widely distributed and additional delays in the circuit. We define two fault models, FDF1 and FDF2. FDF2 is based on two time frames, and FDF1 is more conservative and based on one time frame in the following section. Then we present an ATPG method based on incremental SAT formulations which represents fault lists implicitly. The experimental results are shown next, and the final section gives concluding remarks.

2   Functional Faults Caused by Distributed Additional Delay

As we discussed in the introduction, additional delays due to variation and others can let a gate in a circuit receive possibly incorrect values in the previous cycles rather than the correct ones in the current cycles, which may result in wrong computations by the gate compared with the original functionality of the gate using the correct input values. Please note that the functionality of the gate still remains correct, but the values it uses for computations may be wrong due to additional delays. Let us discuss these issues using an example shown in Fig. 2.

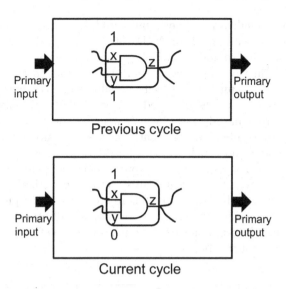

Fig. 2. Values on inputs and output of a gate in a circuit

There is an AND gate in a sequential circuit. For normal operations, the output, z, of the AND gate is 1 for the previous cycle and 0 for the current cycle, as the input values of the gate are (1, 1) and (1, 0) respectively as shown in the figure. Now assume that there are significantly large and distributed additional delays in the circuit due to variation and others. Such delays can let the AND gate get the previous values of the input, x and y, instead of the current ones. Under this situation, the output becomes 1 for the "current" cycle, as the input values the AND gate actually received are (1, 1), i.e., the previous values. This is observed as an incorrect function, which is different from AND operation, as the current inputs are (1, 0) but the output observed is 1. An important observation here is that it is possible that only x-input of the AND gate gets the previous value, which results in the situation where the output of the AND gate is still correct, as the values in the previous cycle and the current cycle for the x-input are the same. On the other hand, if only the y-input of the AND gate gets the values in the previous cycle, the output of the AND gate becomes wrong. It becomes 1 instead of 0 which is correct.

In this section we discuss how functionality of a gate may look like changed due to such delay increase. In general, the value of a signal can be the one for the current or the one in the previous cycle due to delays, and so there are possibly four combinations of values for the current and previous values, i.e., (previous, current) = (0, 0), (0, 1), (1, 0), and (1, 1) for an input of a gate. Obviously if the current and previous values are the same, there will not be any changes in the observed function. So the cases to be examined are the ones where (previous, current) = (0, 1) and (1, 0). Also, depending on the values of the other inputs of the gate, the observed functionality of the gate may or may not change. In order to change the functionality, the other inputs need to be so called non-controlling values, i.e., 0 for OR gate and 1 for AND gate, or those other inputs must also change their values due to additional delays simultaneously.

For simplicity, in this paper we assume that inputs to the combinational part of a sequential circuit can have any possible value combinations. This is, in general, not true, as values provided by the flipflops are only the ones for reachable states from initial states, which may not be all states. Accurately speaking, as we are dealing with sequential circuits, we need to manage which are "reachable" states and which are not in order to precisely compute effects of the additional delays. As reachability computation is very expensive for practical sizes of designs, here we simply assume all states are reachable. This is the same assumption used in scan-based testing. There are ways to compute supersets of reachable states, such as using techniques for property directed reduction [9,10], but utilization of such techniques within our proposed method is a future topic and out of the scope of this paper.

x	y	NoFault	zx1	zx2	zx3	zy1	zy2	zy3	zxy1	zxy2	zxy3
0	0	0	0	0	0	0	0	0	*1*	0	*1*
0	1	0	*1*	0	*1*	0	0	0	0	0	0
1	0	0	0	1	0	*1*	0	*1*	0	0	0
1	1	1	1	*0*	*0*	1	*0*	*0*	1	*0*	*0*
Resulting function	AND	y	$x\tilde{~}y$	$\tilde{~}xy$	x	0	$x\tilde{~}y$	ENOR	0	$\tilde{~}x\tilde{~}y$

Fig. 3. Functionality changes of AND gate due to input delays

Now let us discuss how additional delays can affect the observed functionality of AND and OR gates in the given circuit. Figure 3 shows partial possible behaviors of an AND gate with additional delays. The column, "NoFault" shows the truth table values of the correct AND operation. For each value combination of x and y, that is, for each row of the truth table, the gate may get the previous values of x and/or y instead of the current values if there are delay faults, and those previous values can be different from the current values. The column, "zx1" shows the case where only in the second row of truth table, the AND gate gets the value of x in the previous cycle and that value is 1 which is different from the value in the current cycle. Due to this incorrect value, the output of the AND gate becomes 1 which is wrong as shown with underlined italic in the figure. Assuming that this is the only error in the output of the AND gate, as shown in the truth table, the resulting observed logic function at the output of the AND gate is y instead of $x \wedge y$. Please note the AND gate is performing the correct AND operations, but one of its inputs gets the wrong value. The column, "zx2", shows the case where only the fourth row of the truth table changes its value due to the late arrival of x-input of the AND gate. Here we assume that such late arrival value, which is 0, is different from the correct current value, which is 1. So the resulting observed function becomes $x \wedge \neg y$ instead and simple AND. The column, "zx3" shows the case where these two errors happen simultaneously. As we mentioned above, we assume additional delays in the circuits can happen in distributed and independent ways.

Columns, "zy1","zy2", and "zy3", show the corresponding cases where values of y-input of the AND gate arrive late and their previous and incorrect values are used by the AND gate. As seen from the figure, the resulting observed functions are, x, 0 (constantly 0 function), and $x \wedge \neg y$. Moreover, if values of both x-input and y-input may arrive late, which are the cases shown in columns, "zxy1","zxy2", and "zxy3", as seen from the figure, the resulting observed functions are, $exlusive - nor$, 0 (constantly 0 function), and $\neg x \wedge \neg y$. Figure 3 shows that there are seven incorrect functions possibly observable at the output of the AND gate, if inputs of the gate arrive late and the previous wrong values are used.

Please note that as discussed before, the previous values may or may not. be different from the current correct values. It depends on the behavior of the sequential circuits. The discussions here is assuming the cases where the previous values are different from the current ones.

Similar analysis results are shown in Fig. 4 for an OR gate. Similar to the cases of the AND gate, Fig. 4 shows that there are also seven incorrect functions possibly observable at the output of the OR gate. Columns, "zxy1","zxy2", and "zxy3", show the corresponding cases where the x-input of the OR gate gets the wrong values, and columns, "zy1", "zy2", and "zy3", show the corresponding cases where the y-input of the OR gate gets the wrong values. Columns, "zxy1", "zxy2", and "zxy3", show the corresponding cases where both of the x-input and y-input of the OR gate gets the wrong values,

x	y	NoFault	zx1	zx2	zx3	zy1	zy2	zy3	zxy1	zxy2	zxy3
0	0	0	1	0	1	1	0	1	1	0	1
0	1	1	1	1	1	1	0	0	1	1	1
1	0	1	1	0	0	1	1	1	1	1	1
1	1	1	1	1	1	1	1	1	1	0	0
Resulting function		OR	1	y	~x+y	1	x	x+~y	1	EOR	~x+~y

Fig. 4. Functionality changes of OR gate due to input delays

Please note that here we have analyzed only a subset of possible behaviors. It is possible that the values of x and y can be independently chosen to be the previous values. By observing the truth tables shown in Figs. 3 and 4, we can realize that each row of the truth tables for AND/OR functions could change its value with appropriate late arrival of inputs and different values in the previous cycle from the ones in the current cycle independently. This means that in our models, essentially all possible logic functions with two-inputs can potentially be realized by the delays due to variations and others. Therefore, for the analysis of faulty behaviors caused by late arrival of signal values, all functional faults, which are 15 in total in the case of two-input gates, should be taken into account. Of course this depends on the behaviors of the given sequential circuit. It may realize all of the 15 functions under faults, or it may not. Therefore, it is important to take into account the sequential behaviors in two time frames, the current and the previous cycles.

Please also note that this discussion is true only if we can freely choose the values as the ones for the previous cycles. In real sequential circuits, however, the values in the previous cycles are determined by the sequential circuits themselves and can not be freely chosen. As we will show in the following section, we define two functional delay fault models, one with one time frame, called FDF1, and the other with two time frames, called FDF2. In FDF1 model, we assume the values in the previous cycle can be freely chosen, which means that values for any inputs of gates can become wrong. Here we need just one time frame to define the fault. On the other hand, in FDF2 model, the values in the previous cycle are determined by the sequential circuits, and so we need two time frames for defining the faults.

3   Functional Delay Fault Models

Based on the discussions in the previous section, in this section we present two functional delay fault models, FDF1 and FDF2. FDF2 is the one with two time frames and FDF1 is the one with one time frame. The former is basically following the discussions in the introduction section while the latter is based on a simplified assumption from the former. Please note that in our functional delay fault model, it is essential to deal with "multiple" faults, as additional delays by variations and others are widely distributed in a circuit, and the values in many internal signals may change their values simultaneously. We introduce these functional delay fault models in the rest of this section.

3.1   Functional Delay Fault Model with Two Time Frames, FDF2

Because we need information on the values of signals in the current cycle as well as the ones in the previous cycle, given sequential circuits must be time-frame expanded by two times. For example, an example sequential circuit and its two time-frame expanded one are shown in Figs. 5 and 6 respectively. Please note that although there are flipflops in the expanded circuit shown in Fig. 6, those should be analyzed as pure buffers with no delays for the following mathematical analysis. That is, in our analysis, circuits are considered as pure combinational circuits with no delays in flipflops of the circuits.

Our first fault model caused by additional delays, are called Functional delay fault model with two time frames, FDF2, and it assumes that under faults, the values of the inputs of a number of gates in the circuit are the ones in the previous cycle instead of the current cycle. This can be represented with a multiplexer for each input of a gate in the second time frame of the expanded circuit. The 0-input of the multiplexer is connected to the original source whereas the 1-input is connected to the corresponding signal of the gate in the first time frame of the expanded circuit. Example insertions are shown in Fig. 7. Please note that for

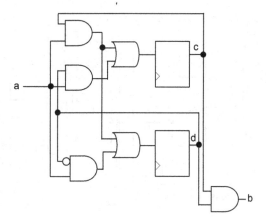

Fig. 5. An example sequential circuit

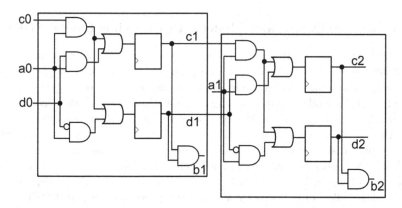

Fig. 6. Two time-frame expanded circuit from Fig. 5

easiness of drawing figures, only one gate in the second time frame is converted to have such multiplexers in its inputs. In actual modeling the inputs of all gates in the second time frame of the expanded circuit should have their multiplexers.

These multiplexers allow the inputs of the gates in the second time frame to get either values in the current cycle or the ones in the previous cycle depending on the control signals, $v1, v2$, of the multiplexer. Those control signals are called parameter variables and represent which faults are active in the circuit. Please note that if both of them are 0, there is no fault on the inputs of that gate. Therefore, if the summation of the numbers of inputs of all gates is m in the combinational part of the given sequential circuit, there are totally $2^m - 1$ multiple fault combinations in the circuit. As we said, it is essential to deal with all of these fault combinations, or as many as possible, when we perform ATPG for functional delay fault testing.

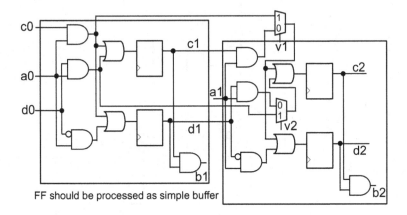

Fig. 7. Multiplexers are added to a gate in the circuit shown in Fig. 6

3.2   Functional Delay Fault Model with One Time Frame, FDF1

The functional delay fault model defined in the previous sub-section uses two time frames as the value of an input of a gate can get the value of the previous cycle rather than the current cycle. This means we need to analyze two time frames for ATPG, and so simpler model with one time frame could become useful if circuits become larger. From the viewpoint of the functions and operations of gates, the faulty values of the previous cycle used in the previous sub-section may be replaced simply with the complemented values of the current ones. Such complemented values are always incorrect, which means the resulting fault model is more conservative, but need only one time frame. This fault model with one time frame is called Functional Delay Fault with one time frame, FDF1, in this paper.

Under faulty situations, this fault always introduces incorrect values to the inputs of the gates in the circuit whereas the fault in the precious sub-section introduces incorrect values only when the values of the current and previous cycles are different. So the fault model defined in this sub-section introduces more erroneous values, and as a consequence, if we completely test given circuits with this fault model, we may be testing too much and so called "over-testing" problem could happen. That is, infeasible situations are also taken into account when generating test vectors. Please note that even the fault model in the previous sub-section may introduce over-testing as we assume all value combinations are feasible as the values of flipflops. This is essentially the same over-testing problem as the one for full scan based designs with stuck-at faults.

This fault model with one time frame can be represented in a similar way as the previous one by using multiplexers as shown in Fig. 8. Please note that the 1-input is connected to the output of an inverter whose input is connected to the original signal. The original signal is connected to the 0-input of the multiplexer as well. Although multiplexers are inserted into the inputs of one

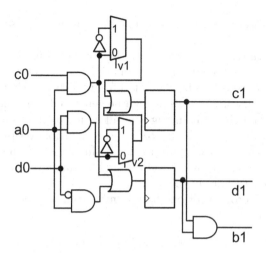

Fig. 8. Multiplexers are added for FDF1 model

gate in the figure for easiness of drawing figures, all inputs of all gates should have multiplexers for multiple faults just like the previous case.

4   ATPG Methods Based on Incremental SAT Formulation

As we assume all states of flipflops are reachable in this paper, the values of pseudo inputs coming from flipflops, that is, inputs, $c0$ and $d0$, in Figs. 6 and 8 are assumed to be able to have all combinations of values. All possible fault combinations under our fault models can be represented by all value combinations of the control inputs of the multiplexers, except for all 0 which represents the non-faulty (fault free) case. Such control inputs are called parameter variables in this paper.

This is an implicit way to represent multiple faults just like the state encoding with state variables in model checking [14]. This method was first proposed in [5]. The number of possible fault combinations is exponential with respect to the number of multiplexers, which is the same as the number of inputs of all gates. With this implicit representation, very large numbers of possible simultaneous faults are represented with exponentially small numbers of variables (parameter variables).

ATPG methods for the two fault models, FDF1 and FDF2, are basically the same. The only difference is how to represent faults with multiplexers and their associated parameter variables. Let x be the set of inputs to the one time frame or two time frame circuit, and v be the set of control signals of multiplexers, that is, parameter variables. Please note that the x variables in the two time frame circuits represent both primary inoputs of the first and second time frames, and the v variables exist only in the second time frame as only the second time frame has multiplexers.

Also, let $NoFault(x)$ and $Faulty(v, x)$ be the logic functions realized at the outputs by the circuit without and with multiplexers, respectively. An example of formula of $NoFault(x)$ can be generated from the circuit shown in Fig. 6, and an example of formula of $Faulty(v, x)$ can be generated from the circuits shown in Figs. 7 and 8 respectively assuming that all inputs of gates have multiplexers.

Although $NoFault(x)$ and $Faulty(v, x)$ are multiple output functions, for easiness of notations, we write them just like a single output function. For example, their equality, i.e., all output values are the same, is simply described as $NoFault(x) = Faulty(v, x)$ in this paper.

Then an ATPG process for one fault combination can be formulated as the following SAT problem:

$$\exists v, x. Faulty(v, x) \neq NoFault(x) \tag{1}$$

Please note that this is a normal SAT problem and says some fault can be detected by some input vector, as under that input vector the two circuits behave differently. Let the solution values of variables, (v, x), be (v_1, x_1) respectively. Now we have found that the fault corresponding to v_1 can be detected by the input, x_1.

In traditional ATPG processes, fault simulators are used for the input vector, x_1, to eliminate all of the detectable faults from the target remaining faults (fault dropping process). In our case, this approach does not work as we are dealing with multiple faults and there are so many possible fault combinations which can never be manipulated explicitly (exponentially many with respect to the numbers of multiple faults). Please remind that multiple faults are essential in order to deal with the faults caused by distributed and additional delays. So the question is how to eliminate faults which are detectable by a test vector "implicitly" not explicitly ?

We formulate the ATPG process as a SAT problem in the following way:

$$\exists v. Faulty(v, x_1) \neq NoFault(x_1)$$

where x_1 is one of the solutions for (1). All the faults corresponding to the values of v, which are the solution of the SAT problem, can be detected by the test vector, x_1. Therefore, in order to eliminate the detected faults by the test vector, x_1, when generating next test vector, we should add the following constraint on top of (1):

$$Fauty(v, x_1) = NoFault(x_1)$$

This constrains that values of v should be the ones which behave correctly with test vector, x_1, that is, undetectable faults.

So the next step of our ATPG process is to solve the following SAT problem:

$$\exists v, x. (Faulty(v, x) \neq NoFault(x))$$
$$\wedge (Faulty(v, x_1) = NoFault(x_1)) \tag{2}$$

where x_1 is the solution of (1) above.

Let the solution values of the variables, (v, x), for (2) be (v_2, x_2) respectively. Then x_2 becomes the second input test vector. It detects some faults which cannot be detected by the previous test vector, x_1.

We keep doing this until there is no more solution. Here we assume that the following SAT problem has a solution

$$\exists v, x.(Faulty(v, x) \neq NoFault(x))$$
$$\wedge (Faulty(v, x_1) = NoFualt(x_1))$$
$$\wedge (Faulty(v, x_2) = NoFault(x_2)) \wedge ...$$
$$\wedge (Faulty(v, x_{n-1}) = NoFault(x_{n-1})) \tag{3}$$

but the following SAT problem has no solution, that is, unsatisfiable,

$$\exists v, x.(faulty(v, x) \neq NoFault(x))$$
$$\wedge (Faulty(v, x_1) = NoFault(x_1))$$
$$\wedge (Fauty(v, x_2) = NoFault(x_2)) \wedge ...$$
$$\wedge (Faulty(v, x_{n-1}) = NoFault(x_{n-1}))$$
$$\wedge (Faulty(v, x_n) = NoFault(x_n)). \tag{4}$$

As (3) has a solution and (4) does not have a solution, the input test vectors, $x_1, x_2, ..., x_n$ can detect all of the detectable faults, as the unsatisfiability of the formula (4) guarantees that there is no more detectable fault. So they become a set of complete test vectors for our multiple fault model exclusive of redundant faults. Please note that redundant faults are automatically excluded from the target faults, as redundant faults have no valid test vectors, which means there is no solution for the SAT problem.

Discussions above can be summarized as the flow shown in Fig. 9. The set in testVectors keeps the set of test vectors accumulated so far. The formula in InConstraints excludes all of the faults which are detectable by the current test vectors. The numbers of test vectors required to detect all faults, or in other words, the performance of the ATPG algorithm depends on how many times the formula (3) becomes satisfiable, i.e., numbers of iterations in the loop of Fig. 9. Please note that each test vector is generated explicitly whereas detectable faults by the current set of test vectors are implicitly and automatically excluded from the target fault combinations.

As can be clearly seen from Fig. 9, the SAT problems to be solved are pure "incremental SAT" problem. The formulae are updated to have more constraints, that is, the following formula is a super set of the previous formulae. Therefore, all learning and backtracks made so far in case-split based SAT solvers, which are common nowadays, are guaranteed to be all valid in the following formulae, and the reasoning in the previous formula can simply be continued, not restarted, in the following formula. In reasoning about the formula (1) above, after some number of backtracks, a SAT solver finds a solution (v_1, x_1). The next formula to be checked is (2) where (v_1, x_1) is not a solution, and so the SAT solver simply

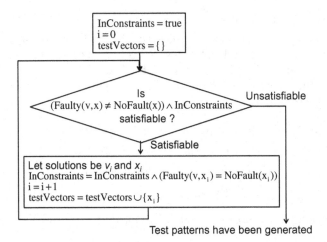

Fig. 9. ATPG flow with incremental SAT

backtracks without any reasoning required. After some number of more back-tracks, the SAT solver finds another solution, (v_2, x_2). This reasoning continues until the expanded formula becomes unsatisfiable, which means case-splitting has covered all cases implicitly.

Now in order to illustrate the ATPG process more clearly, we show an example run for FDF2 testing on a small ISCAS89 circuit, s27 by using an implemented command for the ATPG on top of the logic synthesis and verification tool, ABC [6]. The execution trace on the ABC tool is illustrated in Fig. 10. The implemented command for the proposed ATPG methods for FDF2 faults is "&fftest" with "-A 1" option. The option of "-v" gives detailed execution traces. The command is included in the standard distribution of ABC.

```
abc 03> &r s27.aig
abc 03> &ps
s27     : i/o =     4/    1 ff =    3  and =     8 lev =    5 (3.75)  mem = 0.00 MB
abc 03> &fftest -v -A 1
FFTEST is computing test patterns for delay faults...
Using miter with: AIG nodes =     55. CNF variables =     49. CNF clauses =      103.
Iter    0 : Var =      49 Clause =      103 Conflict =      14
Iter    1 : Var =      77 Clause =      136 Conflict =      14
Iter    2 : Var =      103 Clause =      160 Conflict =      15
Iter    3 : Var =      130 Clause =      189 Conflict =      15
Iter    4 : Var =      157 Clause =      211 Conflict =      15
Iter    5 : Var =      185 Clause =      234 Conflict =      20
Iter    6 : Var =      211 Clause =      239 Conflict =      21
Solver time =    0.00 sec
The problem is UNSAT after 6 iterations. Testing runtime =     0.01 sec
abc 03>
```

Fig. 10. ATPG execution trace example for the benchmark circuit, s27

After reading the circuit, s27, in AIG format, the statistics of the circuit are shown. Then the ATPG command, "&fftest" is invoked. First the formula of (1) above for s27 is solved by a SAT solver. This formula has 103 clauses. After 14 conflicts/backtracks the SAT solver generate a first test vector. In the next step, the formula of (2) above for s27 becomes the target. That has 136 clauses which are 33 more than the previous (first) formula. This additional clauses comes from $(Faulty(v, x_1) = NoFault(x_1))$ part of (2) above as well as the learned clauses in the first SAT solving. In the second run of the SAT solver, it generates second test vector without additional conflict/backtrack. The formula for the third run of the SAT solver has 160 clauses, which are 24 clauses more than the second run, in order to exclude the faults detectable by the second test vector efficiently also with newly learned clauses. The third run finds the third test vector with one additional conflict/backtrack. This process continues and after seven iterations, the resulting SAT formula becomes unsatisfiable. In total six test vectors are generated and the final formula is unsatisfiable. This unsatisfiability can be made sure with 21 conflicts/backtracks in total. That is, the total number of conflicts/backtracks required for all seven (the number of test vectors plus 1 for the final UNSAT problem) SAT solving for s27 is 21. Please note that the final problem is unsatisfiable and needs 21 conflicts/backtracks in total to prove its unsatisfiability for s27.

As can be seen from the above execution trace, the problem is an incremental SAT problem as a whole. Or we can say that we are solving an unsatisfiable problem as a whole, but start with satisfiable ones and add more constraints incrementally based on the test vectors generated. That is, the set of the SAT problems (or formulae) can be considered as a single SAT problem, which should be unsatisfiable eventually. So the overall process of the proposed ATPG method is just to solve single SAT problem to make sure it is unsatisfiable, allowing dynamic addition of more constraints during the SAT reasoning process. That is, each time we find a new test vector, new constraints which exclude the faults detectable by that test vector are added. The learned clauses in the previous run are also included. By slightly modifying existing (case-split based) SAT solvers, we can realize the proposed ATPG method inside SAT solvers.

One remark in our formulation is that ATPG for single, or double, or triple faults, and so on, can easily be formulated within our SAT based ATPG with implicit representations of fault lists. We can add constraints to restrict how many parameter variables can be simultaneously one. If only one parameter variable can be one at a time, it is an ATPG for single faults. In the experiments below, we compare the numbers of test vectors for complete multiple faults (there are $2^m - 1$ fault combinations where m is the number of potential faulty locations) and single faults.

The above discussions can also be casted to non-SAT based ATPG techniques with learning, such as [3, 4], if we introduce additional circuits with parameter variables to represent faults. As ATPG tools are well developed utilizing various circuit-related and structural techniques and reasoning, such ATPG tools with the above method for the representation of detectable faults as circuits can

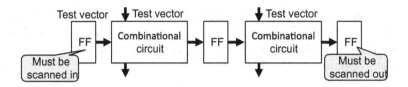

Fig. 11. Scan-based testing for the proposed methods

potentially realize very efficient ATPG tools for our fault models as well. This will be one of our future directions.

4.1   Application of Test Vectors

The generated test vectors for the functional delay fault model, FDF1, are applied to the manufactured chips just like the ones for scan based designs for, say, stuck-at faults, as the test vectors for FDF1 have only one time frame. As for the test voters for the functional delay fault model, FDF2, using the scan chains, they are applied to the manufactured chips in the way shown in Fig. 11.

The test vectors for FDF2 have two time frames. A test vector consists of the values for the flipflops in the first time frame and the values for the primary inputs for the first and second time frames. So the values for the flipflops for the first time frame is scanned in and then the chip runs for two cycles instead of one cycle. Please note that the values for the flipflops in the second time frame are generated inside the chip. After running the chip for two cycles, the values for the flipflops are scanned out. So we do not need any additional mechanisms when applying the test vectors for FDF2, and we can simply use the existing scan mechanisms.

5   Experimental Results

We have implemented the proposed ATPG methods for the proposed functional delay faults, FDF1 and FDF2 on top of ABC tool [6] including the use of previously learned clauses in later SAT solving. For easiness of experiments, all ISCAS89 circuits are first converted into AIG (AND Inverter Graph) format where there are only two-input AND gates and inverters. So all the faults of FDF1 and FDF2 are defined on inputs of those two-input AND gates. The results for FDF2 (functional delay fault with two time frames) are shown in Table 1 and the ones for FDF1 (flip fault with one time frame) are shown in Table 2. One test vector for FDF2 consists of two time frames whereas the one for FDF1 has only one time frame. In both tables, Name is the name of an ISCAS89 benchmark circuit, and PI/PO/FF/AND are the numbers of primary inputs, outputs, flipflops, and AIG nodes used to represent the circuits. Vars/Clauses/Conflicts are the numbers of SAT variables, clauses, and conflicts, and Tests is the total number of test vectors computed using the proposed ATPG algorithm.

Please note that for all circuits with multiple faults of either FDF1 or FDF2, the ATPG processes have finished completely, that is, these sets of test vectors can detect all combinations of the multiple faults as long as they are not redundant. The sets of test vectors detect exponentially many combinations of multiple faults. For large ISCAS89 circuits, there are more than 10,000 AND gates in AIG format. So the numbers of multiple fault combinations are in the order of $2^{10,000}$. Time is the processing time on a sever computer having Linux kernel 2.6.32 64-bit, Dual Xeon E5-2690 2.9 GHz, 128 GB memory.

As seen from the tables, we have succeeded in generating complete test vectors for all multiple faults of FDF1 and FDF2. Redundant faults are automatically excluded from the target faults as unsatisfiable cases. In general as functional delay fault models, FDF2 is more accurate than FDF1, because FDF2 regards the cases where the current and previous values are the same to be automatically non-faulty. The numbers of test vectors and execution times for FDF1 and FDF2 are somehow similar although the numbers of test vectors for FDF1 are slightly smaller for large circuits. On the other hand, the ATPG times for FDF2 is slightly shorter than the ones for FDF1 for large circuits. This may not be intuitively understood as FDF2 needs two time frames whereas FDF needs only one time frame. These are issues for future research with more detailed and intensive experiments.

In both fault models, a couple of thousands of test vectors or less are sufficient to detect all multiple faults on ISCAS89 circuits. This suggests that FDF1 and FDF2 could be reasonable functional delay fault models in practice. FDF2 is better as it works with two time frames and its model is more accurate than FDF1. Although FDF2 does not measure any actual delays of any paths, it tries to cover all possible resulting functionally different cases due to distributed and additional delays. It can detects all possible functional effects caused by delay faults, and so the complete sets of test vectors for all multiple fault combinations may make sense in practice especially with the fact that the numbers of test vectors are not so many as seen from the experimental results.

Finally as for comparison of ATPGs for multiple faults and simple faults, we have also generated complete test vectors for "single" functional delay faults. As discussed above, we should expect wide and distributed delays in circuits, and it makes much more sense for multiple faults rather than single faults. So these results are just to see how many "more" test vectors required and how much more difficult for multiple faults over single faults. As discussed above, it is easy to set constraints for single faults in our formulation, i.e., add clauses to let only one parameter variable be one when generating test vectors.

In order to save the space in the paper, experimental results for large ISCAS89 circuits are compared. The comparisons are shown in Table 3 for FDF2 faults and in Table 4 for FDF1. From Table 3, we can say that for FDF2, the problem sizes in terms of numbers of variables and clauses are 2–3 times difference, whereas the numbers of conflicts/backtracks are around the same (actually a little bit smaller in many cases). This is quite interesting in that for FDF2 model, ATPG for single faults and multiple faults are not much different in terms

Table 1. Complete test generation results for FDF2 on ISCAS89 circuits

Name	PI	PO	FF	AND	Vars	Clauses	Conflicts	Tests	Time (s)
s27	4	1	3	8	211	239	21	6	0.01
s208.1	10	1	8	72	3504	4132	279	27	0.05
s298	3	6	14	102	6172	6994	285	37	0.02
s344	9	11	15	105	7666	8966	286	40	0.02
s349	9	11	15	109	7173	8660	274	35	0.02
s382	3	6	21	140	10801	12571	401	48	0.03
s386	7	7	6	166	16792	20352	435	76	0.05
s400	3	6	21	148	7739	10362	507	32	0.03
s420.1	18	1	16	160	18285	23217	905	71	0.13
s444	3	6	21	155	9071	10717	601	37	0.03
s510	19	7	6	213	19598	17235	691	65	0.05
s526	3	6	21	203	13593	17004	1184	46	0.09
s641	35	24	19	146	24355	19125	589	80	0.05
s713	35	23	19	160	26792	22872	629	84	0.06
s820	18	19	5	345	52472	52085	1096	117	0.16
s832	18	19	5	356	55846	55701	1165	122	0.19
s838.1	34	1	32	336	55812	53108	1446	110	0.56
s953	16	23	29	347	60703	49819	1508	121	0.17
s1196	14	14	18	477	104047	97072	2256	172	0.95
s1238	14	14	18	532	132631	120833	2151	202	0.89
s1423	17	5	74	462	118481	160938	2358	156	0.5
s1488	8	19	6	663	143999	163873	1582	184	0.53
s1494	8	19	6	673	138016	140575	1437	175	0.45
s5378	35	49	179	1389	704511	763815	4486	334	5.96
s9234	19	22	228	1958	1744224	1699261	7356	639	36.75
s13207	31	121	669	2719	4207211	3751671	8757	919	201.17
s15850	14	87	597	3560	4456945	3833199	16824	860	243.74
s35932	35	320	1728	11948	12764553	12765854	18600	719	912.98
s38417	28	106	1636	9219	27676324	26501629	49582	1948	3363.42
s38584	12	278	1452	12400	63168314	61442523	42009	3819	26221.71

of computing complexity. We need more detailed experiments to confirm this, which we are working on.

On the other hand from Table 4, we can say that for FDF1, multiple faults are much more difficult than single faults as the former needs a lot more conflicts/backtracks. However, the numbers of test vectors are different up to 3–4 times or so. The different behaviors in the two functional fault models, FDF1

Table 2. Complete test generation results for FDF1 on ISCAS89 circuits

Name	PI	PO	FF	AND	Vars	Clauses	Conflicts	Tests	Time (s)
s27	4	1	3	8	187	309	24	6	0.01
s208.1	10	1	8	72	10570	22936	1162	76	0.23
s298	3	6	14	102	7271	16662	827	35	0.04
s344	9	11	15	105	7226	17594	2354	30	0.2
s349	9	11	15	109	8234	18308	730	35	0.13
s382	3	6	21	140	12134	26631	918	43	0.17
s386	7	7	6	166	14208	30735	2772	54	0.21
s400	3	6	21	148	12418	28691	1517	41	0.13
s420.1	18	1	16	160	62279	141652	17761	210	2.88
s444	3	6	21	155	11554	28741	2216	36	0.1
s510	19	7	6	213	25663	68806	11058	66	0.52
s526	3	6	21	203	29752	76137	4523	77	0.32
s641	35	24	19	146	27264	55077	1897	80	0.15
s713	35	23	19	160	21224	46485	1639	56	0.13
s820	18	19	5	345	69613	168271	12831	127	1.15
s832	18	19	5	356	89826	217392	16966	158	1.84
s838.1	34	1	32	336	337602	735817	476696	555	368.42
s953	16	23	29	347	68247	172198	19507	94	1.74
s1196	14	14	18	477	127849	312047	16891	155	2.79
s1238	14	14	18	532	173882	419099	18423	194	2.95
s1423	17	5	74	462	88218	193894	9134	90	1.14
s1488	8	19	6	663	122603	274328	12370	131	1.98
s1494	8	19	6	673	138915	316904	18147	148	3.13
s5378	35	49	179	1389	732113	1656343	2101830	263	547.32
s9234	19	22	228	1958	1585886	3570590	5564626	423	2100.66
s13207	31	121	669	2719	2723282	4987122	711130	474	557.43
s15850	14	87	597	3560	3203371	6933658	18862813	440	3625.81
s35932	35	320	1728	11948	3447907	6548092	1070785	173	7709.4
s38417	28	106	1636	9219	17657911	35695772	140866184	901	70512.05
s38584	12	278	1452	12400	14291557	28562060	1298930	609	6095.87

and FDF2, for single and multiple faults may come from the fact that in FDF1 all faults actually introduce wrong values to the circuits whereas in FDF2 even under faulty, the values can still be correct if the values in the previous cycle are the same as the current ones. So the numbers of wrong behaviors introduced to the circuits could be a lot different. This could be part of the reasons, although things are not so sure and need much more experiments. Also, please

Table 3. Comparison of ATPG for single and multiple FDF2 faults

Name	FDF1 single faults					FDF1 multiple faults (normalized with single = 1)				
	Vars	Clauses	Conflicts	Tests	Time (s)	Vars	Clauses	Conflicts	Tests	Time
s5378	327012	817209	6696	116	9.17	2.24	2.03	313.89	2.27	59.69
s9234	500987	1309285	9605	131	28.26	3.17	2.73	579.35	3.23	74.33
s13207	614918	1318897	11706	102	35.92	4.43	3.78	60.75	4.65	15.52
s15850	1102997	2631088	19302	149	77.29	2.90	2.64	977.25	2.95	46.91
s35932	7875486	15535384	40238	373	1535.48	0.44	0.42	26.61	0.46	5.02
s38417	5313144	12680228	91227	269	1472.31	3.32	2.82	1544.13	3.35	47.89
s38584	9331642	22354257	53863	392	2157.43	1.53	1.28	24.12	1.55	2.83

Table 4. Comparison of ATPG for single and multiple FDF1 faults

Name	FDF2 single faults					FDF2 multiple faults (normalized with single = 1)				
	Vars	Clauses	Conflicts	Tests	Time (s)	Vars	Clauses	Conflicts	Tests	Time
s5378	530545	576418	4795	249	2.84	1.33	1.33	0.94	1.34	2.10
s9234	1125880	1215152	7690	414	17.88	1.55	1.40	0.96	1.54	2.06
s13207	2731864	2187226	8033	599	62.76	1.54	1.72	1.09	1.53	3.21
s15850	2888242	2472141	20667	559	100.43	1.54	1.55	0.81	1.54	2.43
s35932	9495349	8829489	23100	527	411.34	1.34	1.45	0.81	1.36	2.22
s38417	13426351	11950943	53452	943	689.76	2.06	2.22	0.93	2.07	4.88
s38584	24822660	16479553	41426	1496	2534.52	2.54	3.73	1.01	2.55	10.35

note that we did not spend any efforts to try to make the sets of test vectors more compact. Instead we just solve the incremental SAT problems. We do need to analyze much more details with intensive experiments, but the tables shown in this paper can give a good first step.

6   Concluding Remarks

We have shown functional delay fault models caused by delay variations and their associated ATPG methods with implicit representations of multiple fault lists. We are recognizing that the algorithm shown in Fig. 9 is essentially doing the same or very similar as the techniques introduced in [7,8], although the goals are different. Similar ATPG methods have been developed targeting multiple stuck-at faults [5].

As discussed in the literature, the problem to be solved is naturally formulated as QBF (Quantified Boolean Formula), but solved through repeated application of SAT solvers, which was first discussed under FPGA synthesis in [11] and in program synthesis in [12]. [13] discusses the general framework on how to deal with QBF only with SAT solvers.

The largest ISCAS89 circuits have more than ten thousands two-inpur AND gates, which means that there are more than $2^{(ten\ thousands)}$ of multiple fault combinations. For such large numbers of fault combinations, according to our experiments, a couple of thousands of test vectors are sufficient to detect all of them exclusive of redundant faults. This is a very important and also interesting result, as our functional delay fault models make good sense if we can deal with wide varieties of multiple faults. This is because the effects of additional delays can be distributed very widely, and as a result, there can be many simultaneous value errors happening in the circuit.

Also, future directions include comparison with multiple stuck-at faults from the viewpoints of test vectors, such as the test vectors for our fault models can detect how much of multiple stuck-at faults and vice versa. As we can deal with large numbers of multiple faults, application of the proposed techniques to functional design verification should also be included as future directions.

Finally we like to mention about over-testing issues. As we assume flipflops can have all combinations of values, test vectors may be generated using "unreachable" states. This is a general problem for ATPG with scan based designs. In formal verification fields, there have been significant works performed on computing approximate reachable states or smallest supersets of reachable states. It may be interesting to see how test vectors are affected with constraints coming from such reachable/unreachable states. Given sets of approximated reachable states are simply added to our formulation as additional constraints. Another important issue on the numbers of test vectors is their compaction. There have been works on test vector compaction with SAT-based ATPG, such as [15]. How we can utilize such techniques on compaction with our multiple fault ATPG is clearly one of the very important future researches.

References

1. Qiu, W., Walker, D.M.H.: An efficient algorithm for finding the k longest testable paths through each gate in a combinational circuit. In: IEEE International Test Coference (ITC), pp. 592–601 (2003)
2. Sauer, M., Kupferschmid, S., Czutro, A., Polian, I., Reddy, S.M., Becker, B.: Functional test of small-delay faults using SAT and Craig interpolation. In: IEEE International Test Coference (ITC) (2012)
3. Schulz, M.H., Trischler, E., Sarfert, T.M.: SOCRATES: a highly efficient automatic test generation system. In: IEEE Transaction on Computer Aided Design, pp. 126–137, January 1988
4. Giraldi, J., Bushnell, M.L.: Search State Equivalence for Redundancy Identification and Test Generation, International Test Conference (ITC), pp. 184–193 (1991)
5. Fujita, M., Mishchenko, A.: Efficient SAT-based ATPG techniques for all multiple stuck-at faults. In: International Test Conference (ITC) (2014)
6. Brayton, R., Mishchenko, A.: ABC: an academic industrial-strength verification tool. In: Touili, T., Cook, B., Jackson, P. (eds.) CAV 2010. LNCS, vol. 6174, pp. 24–40. Springer, Heidelberg (2010)
7. Jo, S., Matsumoto, T., Fujita, M.: SAT-BEfficient implementation of property directed reachability. In: Formal Asian Test Symposium (ATS), pp. 19–24, November 2012
8. Fujita, M., Jo, S., Ono, S., Matsumoto, T.: Partial synthesis through sampling with and without specification. In: International Conference on Computer Aided Design (ICCAD), pp. 787–794, November 2013

9. Bradley, A.R.: SAT-based model checking without unrolling. In: Jhala, R., Schmidt, D. (eds.) VMCAI 2011. LNCS, vol. 6538, pp. 70–87. Springer, Heidelberg (2011)
10. En,N., Mishchenko, A., Brayton, R.K.: Efficient implementation of property directed reachability. In: Formal Methods in Computer-Aided Design (FMCAD) (2011)
11. Ling, A., Singh, D.P., Brown, S.D.: FPGA logic synthesis using quantified Boolean satisfiability. In: Bacchus, F., Walsh, T. (eds.) SAT 2005. LNCS, vol. 3569, pp. 444–450. Springer, Heidelberg (2005)
12. Solar-Lezama, A., Tancau, L., Bodik, R., Seshia, S.A., Saraswat, V.A.: Combinatorial sketching for finite programs. ASPLOS **2006**, 404–415 (2006)
13. Janota, M., Klieber, W., Marques-Silva, J., Clarke, E.: Solving QBF with counterexample guided refinement. In: Cimatti, A., Sebastiani, R. (eds.) SAT 2012. LNCS, vol. 7317, pp. 114–128. Springer, Heidelberg (2012)
14. Clarke, E.M., Klieber, W., Nováček, M., Zuliani, P.: Model checking and the state explosion problem. In: Meyer, B., Nordio, M. (eds.) LASER 2011. LNCS, vol. 7682, pp. 1–30. Springer, Heidelberg (2012)
15. Eggersglus, S., Wille, R., Drechsler, R.: Improved SAT-based ATPG: more constraints, better compaction. In: International Conference on Computer Aided Design (ICCAD), pp. 85–90 (2013)

A Temperature-Aware Battery Cycle Life Model for Different Battery Chemistries

Alberto Bocca[1]([✉]), Alessandro Sassone[1], Donghwa Shin[2], Alberto Macii[1], Enrico Macii[1], and Massimo Poncino[1]

[1] Politecnico di Torino, Corso Duca Degli Abruzzi 24, 10129 Turin, Italy
{alberto.bocca,alessandro.sassone,alberto.macii,
enrico.macii,massimo.poncino}@polito.it
[2] Yeungnam University, 280 Daehak-Ro,
Gyeongsan, Gyeongbuk 712-749, Republic of Korea
donghwashin@yu.ac.kr

Abstract. With the remarkable recent rise in the production of battery-powered devices, their reliability analysis cannot disregard the assessment of battery life. In the literature, there are several battery cycle life models that exhibit a generic trade-off between generality and accuracy.

In this work we propose a compact cycle life model for batteries of different chemistries. Model parameters are obtained by fitting the curve based on information reported in datasheets, and can be adapted to the quantity and type of available data. Furthermore, we extend the basic model by including some derating factors when considering temperature and current rate as stress factors in cycle life.

Applying the model to various commercial batteries yields an average estimation error, in terms of the number of cycles, generally smaller than 10 %. This is consistent with the typical tolerance provided in the datasheets.

Keywords: Battery modeling · Cycle life · Battery chemistry · Capacity fading

1 Introduction

Rechargeable batteries are an essential component in many application domains, such as electric vehicles, mobile systems, renewable energy, and telecommunication systems. In order to carry out an early verification of these systems, including the exchange of energy between the energy storage devices and other components, it becomes essential to have accurate and efficient battery models, especially models that evaluate the lifetime of the battery in terms of useful charge-discharge cycles.

In the literature various models for different functional aspects of batteries have been proposed, with differing tradeoffs between accuracy and generality. In the field of electronic design, the most commonly used ones are those in

Y. Shin et al. (Eds.): VLSI-SoC 2015, IFIP AICT 483, pp. 109–130, 2016.
DOI: 10.1007/978-3-319-46097-0_6

which the battery is described by a generic standard model expressed in terms of an *equivalent electrical circuit*. (e.g., [1,2]). This is then populated either using data obtained from direct measurements on actual devices or by extrapolation of battery characteristics available from datasheets (e.g., [3]). These kinds of models are typically generated for a specific battery chemistry and show a high degree of accuracy. This accuracy may significantly degrade if these models are applied to different battery chemistries. Furthermore, they are specific to a given battery chemistry and thus show a very high degree of accuracy. Obviously, this degree of accuracy can vary (decrease) significantly if the model, generated for a particular battery chemistry, is applied to batteries with different chemical characteristics.

On the other hand, in certain contexts (e.g., automotive, aerospace, smart grids), designers often rely on simpler compact analytical macromodels, such as Peukert's law [4], as a quick estimator for the sizing of the battery sub-system or for preliminary what-if analysis. These macromodels are aimed at the generation of a general relationship between the battery *intra-cycle* runtime and the most relevant parameters, like the Depth of Discharge (DOD) or State of Charge (SOC) of a battery.

While these models have reasonable generality (e.g., they can be applied to various batteries with different chemical characteristics, once characterized), they are focused on a single charge/discharge cycle of a battery. They do not provide information about the "lifetime" of a battery, i.e., decrease in performance due to long-term *inter-cycle* effects, such as the fading of the total capacity (ampere-hour) caused by repeated cycling. It is possible to incorporate such aging effects into these circuit-level or analytical models, for instance by replacing the use of a fixed battery capacity value with a generic function of some parameters. However, this operation requires (i) an understanding of the various phenomena that affect battery aging, and (ii) the construction of a compact model that can be used either as a standalone model or incorporated in traditional functional battery models.

The literature provides several studies on these effects, proposing mathematical models that are based on the electrochemical properties or the physics of the batteries and are therefore strongly bound to specific battery materials and chemistry (e.g., [5–9]). Although some other aging models, such as those proposed in [10–15], are empirically characterized onto a pre-defined equation template, they are still derived by measurements and, therefore, are not general enough to support different battery chemistries.

The objective of this work is the generation of an aging model with similar characteristics to a Peukert-like equation. This should be (i) analytical, but able to be empirically populated, and (ii) general enough to support different battery chemistries. Specifically, we propose a mathematical model for estimating the number of cycles with respect to the related capacity fade of batteries.

The accuracy of the approach proposed is demonstrated by applying this model to various commercial batteries of different chemistries, for which the manufacturers provide information on the long-term effects in their datasheets.

The results show an average estimation error, referring to the number of cycles, generally within 10 %, which is consistent with the typical tolerance provided in various datasheets (e.g., [16]).

The paper is organized as follows. Section 2 reports related works on battery modeling, while Sect. 3 describes the proposed mathematical model for estimating the number of cycles of batteries, and Sect. 4 reports the experimental results. In addition, Sect. 5 reports the proposed model extended to the temperature and current effects on battery aging, with the related results, while Sect. 6 draws some conclusions.

2   Background and Motivations

2.1   Battery Aging Issues

The life degradation of a rechargeable battery depends on some irreversible changes of physical, mechanical, and chemical nature (e.g., [17,18] for lithium-ion batteries) in its basic components, such as (i) corrosion, cracking, plating, or exfoliation of the electrodes, (ii) decomposition of the electrolyte and/or of the binder, and (iii) corrosion of the separator, just to list the most evident ones.

The most tangible effect of such deterioration is the irreversible reduction of the total battery capacity, which is named *capacity fade*. This fading in capacity is often measured by the so-called state-of-health (SOH), calculated as the ratio between the actual total capacity C_{aged} and the rated capacity C_R (i.e., the total capacity of one fresh battery), as reported in (1), while the difference C_R - C_{aged} defines the capacity loss (i.e., C_{fade}). In this case, most manufacturers provide information on fading as a percentage (i.e., in a normalized form).

$$SOH = \frac{C_{aged}}{C_R} \tag{1}$$

Battery aging is largely determined by:

- **Temperature (T)**. As with other typical reliability mechanisms, aging usually increases with increasing temperatures; as energy generation process in the battery involves a chemical reaction, the relation with temperature follows an Arrenhius-type of equation. Section 5.1 describes the main temperature effects on cycle life, from a battery perfomance point of view.
- **Depth-of-Discharge (DOD)**. The DOD is the percentage of battery capacity that has been discharged before starting a new charge phase. A DOD of 100 % implies that a battery has been fully discharged before starting a new charge phase. Aging increases with deeper discharge cycles (i.e., higher DOD values).
- **Charge/discharge current**. Both currents affect battery degradation, but generally with a different impact on aging (e.g., [19]). Aging worsens with larger charge or discharge currents. Impact of a certain current on aging strictly depends on the battery chemistry and temperature. Section 5.2 faces this issue considering an analysis for various batteries.

- **Number of cycles (N)**. In a given cycle, deterioration mainly depends on the working and operating conditions. In addition, it may also depend on the number of charge/discharge cycles previously encountered or, in other terms, on the battery SOH at which a certain cycle is performed.

2.2   Battery Aging Models

Although various models have been proposed in literature, they usually have many parameters whose values have to be empirically extracted from direct analysis. For instance, [5] proposes an aging model for a certain lithium-ion (Li-ion) battery that relies on crack propagation theory, with some battery specific constants also related to mechanical strain. It further includes the average state-of-charge (SOC) in the model, since battery aging generally increases for high average SOC values. However, although that mathematical comprehensive model is well-known in the literature, there are practical difficulties to adapt it to different Li-ion batteries.

Concerning cycle life estimation, numerous researchers have proposed analytical models capturing the main aging mechanisms and capacity fading based on the electrochemical properties of the batteries and even including full-physics based models (e.g., [8] for Li-ion batteries). In fact, the causes for degradation in batteries generally differs when considering the various cell components (e.g., electrolyte chemical composition, electrodes design, and active material) [17]. However, from the perspective of an electronic designer this modeling approach is unfeasible and, therefore, more simple and generic aging models are searched. In this work, we focus on compact mathematical battery cycle life models with only a couple of parameters in their formulas, other than the aforementioned aging factors (e.g., DOD and N).

In [9] the authors proposed a model to calculate the usable number of cycles N of a battery based on the following equation:

$$N = N_1 \cdot e^{\alpha \cdot (1 - DOD')} \tag{2}$$

where DOD' is the normalized depth of discharge ($0 \leq DOD' \leq 1$), α is a characteristic constant of the battery and N_1 is the number of cycles at $DOD' = 1$. This model is empirically characterized for lead-acid, nickel-cadmium (NiCd) and nickel-metal hydride (Ni-MH) batteries, whose cycle-life vs. DOD curve has an exponential shape. It is not, however, suitable for many lithium-based cells, whose cycle-life vs. DOD curve sometimes exhibits a more linear behavior (e.g., for $LiFePO_4$ cells).

A slightly different relationship between cycle-life and DOD was introduced in [10]:

$$N = N_{0.8} \cdot DOD' \cdot e^{\alpha \cdot (1 - DOD')} \tag{3}$$

where $N_{0.8}$ is the cycle life at $DOD = 80\%$, while α is a constant whose value is, respectively, 3 and 2.25 for lead-acid and Ni-MH tested battery packs.

Thaller [11] has defined another relationship for battery cycle life after considering excess capacity F, with respect to the rated capacity, and a penalty factor due to the DOD, by including the P parameter, as reported in (4), which gives this mathematical prediction model for a general battery:

$$N = \frac{1 + F - DOD'}{A \cdot (1 + P \cdot DOD') \cdot DOD'} \tag{4}$$

In our work, F is always considered equal to 0, so that each analysis is performed after starting from the rated capacity of any commercial cell or cell string. The product $A \cdot DOD'$ represents the irreversible capacity loss in each cycle. Values of the parameter A were originally declared to be in the range 0.000 $\div$ 0.002 [11].

These previous models estimate the cycle life of a battery, always after considering a fixed irreversible capacity fading (e.g., 20 %, that is, when the total maximum available capacity reaches 80 % of the nominal one).

In [12] the authors introduce a complex cycle life model consisting of different equations, one for each stress factor considered, i.e., C-rate, T and DOD. Despite its high accuracy, the model derivation requires extensive empirical measurements and the model itself lacks the compactness and the generality of a Peukert-like equation.

Another analytical method for battery life prediction is based on the *ampere-hour throughput*, i.e., the total energy supplied by the battery during its life [13], also called "charge life". The charge life Γ_R in ampere-hours (Ah) is defined as:

$$\Gamma_R = L_R \cdot DOD' \cdot C_R \tag{5}$$

where C_R is the rated capacity in Ah at a rated discharge current I_R, and L_R is the maximum number of cycles referring to a given normalized depth of discharge DOD' and a discharge current I_R. In the model presented in [14], the authors proposed calculating an equivalent Ah weighted-throughput parameter.

The model proposed in [15] adopted this approach to estimate the cycling capacity fade through a modified definition of the Arrhenius equation, characterized by a square root time dependence.

2.3 Motivations for the Work

Nowadays, with the remarkable rise in the production of battery-powered electronic devices, system-level design requires an analysis of both circuit and power supply in order to optimize the entire system [1]. Furthermore, battery technology is always "work in progress", as novel battery chemistries are continuously proposed. For instance, during the last two decades Li-ion batteries have mostly replaced NiCd and Ni-MH batteries in mobile phones and portable computers, mainly due to a greater specific energy (Wh/kg) [20].

Therefore, although various models have been proposed in the literature for specific battery types, a more general and flexible model for different chemistries, but still simple enough for fast characterization and simulation, is required.

In spite of the various differences, all the aforementioned models reported in Sect. 2.2 are built by extracting parameter values through measurements on

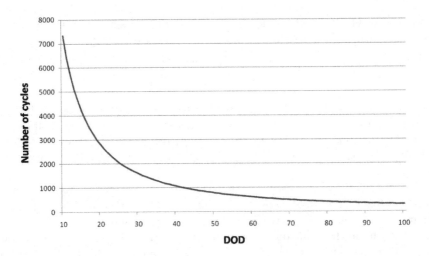

Fig. 1. A typical plot of *Number of cycles vs. DOD.*

the batteries under test. Although the generated models are typically very accurate, this approach is quite time-consuming (especially when multiple cycles are involved) and requires expensive laboratory instrumentation.

There are other methods for analyzing cycle life through computer simulation [21], but they consider the complex governing equations of the chemical reactions. For this reason, methods that only rely on available manufacturer data (e.g., datasheets) to derive the capacity fade in batteries using analytical models (e.g., [22]) or equivalent electrical circuits (e.g., [23]) have been reported in the literature in recent years. Clearly, the accuracy of these models depends on the amount of available information reported in battery datasheets.

The main result of this work is to provide a compact model [24], which expresses *the number of usable cycles as a function of the DOD*, extended for including the other factors affecting capacity fade, namely temperature and charge/discharge current.

The basic outcome of the characterization is a N vs. DOD curve, such as the one shown in Fig. 1. This information is seldom available in typical datasheets and has to be extracted by building an analytical model according to the methodology described in the next section. Needless to say, for the rare cases in which this information is available in the datasheet, the plot can be used directly without resorting to our method. However, in this work we will also consider batteries whose datasheets provide this information, in order to validate our proposed model.

3   Modeling Methodology

3.1   Model Definition

The model proposed in this work somehow mimicks the shape of Peukert's law, as expressed by (6), which models the *intra-cycle* non-linear dependency between capacity and the discharge current:

$$t = \frac{C}{I^k} \tag{6}$$

where C is the capacity of the battery, I is the discharge current, and t is the time for totally discharging the battery; k is the Peukert coefficient; typical values of k depend on the battery chemistry and the manufacturing process and they typically range from 1.1 to 1.3. As a matter of fact, the curves describing the *Capacity vs. Number of cycles* exhibit a similar non-linear relationship.

Our objective is therefore to derive a model expressing battery cycle life in a compact mathematical form similar to Peukert's law, and describing the general non-linear relationship between the capacity fade and the DOD.

In the case of capacity fade, the non-linearity concerns both the number of cycles N as well as the DOD, and the actual relationship among these quantities depends also on the value of the target capacity degradation (i.e., the behavior for a 20 % capacity fade will be different from that for a 30 % capacity fade). In order to model this non-linearity we need to define a new parameter that characterizes the battery performance during the cycling.

The proposed mathematical model is shown in (7); it allows to estimate the number of charging-discharging cycles N for a given battery based on four main parameters.

$$N = L \cdot \frac{C_{fade}}{DOD^h} \tag{7}$$

- L (called the *empirical factor*) is the parameter that is used to calibrate the second term of the model with respect to the number of cycles.
- C_{fade} is the percentage of irreversible capacity loss for which battery life: usually it is considered as 20 %, but some manufacturers considers a different value (e.g., 30 %).
- DOD is the depth of discharge expressed as a percentage (eg. 50 %); to avoid division by 0, it must be > 0, so its range is 1–100 %.
- h is the coefficient that models the nonlinear relationship between N and DOD for a certain C_{fade}.

The similarity with Peukert's law is evident. N, considered as an *inter-cycle* "lifetime" parameter, is obtained as the ratio of capacity fade and a weighted metric of the rated capacity discharged on average per cycle (DOD^h). There are however two relevant differences: (i) factor L is used to scale the "lifetime" across multiple cycles, and (ii) h is not constant, but depends on C_{fade}. This makes our approach more general with respect to previous models and allows one to adapt it to the available manufacturer's data. In fact, the proposed model have two degree of freedom, i.e., L and h, while in the aforementioned models in Eqs. (2), (3), and (4) reported in Sect. 2, one of the two parameters is always fixed because it is strictly related to a physical characteristic, while only the other might be set in order to fit the cycle life function.

Concerning the typical range of DOD, most manufacturers avoid using very low values of DOD (which will results in very large values of N, besides being unrealistic) and usually provide data for DOD in the range from 10–30 % to 80–100 % [25]. Moreover, in case of only a few cycles in a long period of time, aging is usually more influenced by calendar life than cycle life.

The model of Eq. (7), by defining a generic model template, is adaptable also to some batteries like some LiFePO$_4$ batteries, which report a strictly linear *Capacity vs. Number of cycles* characteristic; for this battery, a value of h closer to 1 will fit easily the linear dependency.

3.2 Analysis of the Mathematical Model

In Eq. (7), C_{fade} is constant, and fixed to a standard value, i.e., 20 % as in typical datasheets. Besides the "physical" quantities (C_{fade} and DOD), the model includes two other scale parameters, i.e., the empirical factor L and the binding coefficient h, which have to be determined by fitting empirical data derived from available information (e.g., datasheet). These two parameters reflect a specific characteristic of the battery behavior during its cycle life.

The empirical factor L usually has a value with an order of magnitude comparable to the value of N at low (e.g., 10 or 20 %) DODs. In other words, we can see L as a factor that calibrates the value of the second term of the model (the fraction). Since C_{fade} is constant for a given battery, the fraction actually reduces to $1/DOD$ ^{h}. By plotting this expression as a function of the DOD (Fig. 2) for different values of h, we can clearly see how the non-linearity of $1/DOD^h$ is modulated quite markedly by h. For large (≥ 1) values of h, the curve tends to flatten out, implying that the fraction $1/DOD^h$ tends to become independent of DOD, and relatively low (<0.1). Smaller values of h, conversely, emphasize the dependency on DOD, resulting in significant differences (in order of 0.15–0.2) between low and high DOD values.

The analysis also implies that it is not possible to extract this factor only by analyzing the battery *inter-cycle* behavior, so an algorithm should be run in order to find the two parameters L and h generating the model that best fits the battery cycle life characteristic.

In the next section we present such an algorithm, which searches for the values of both L and h that populate the model having the minimum error in the cycle life estimation with respect to the actual data.

3.3 Extraction of Model Parameters

The actual parameter identification depends on the amount of available data. Many manufacturers provide information about capacity fade in the form of a

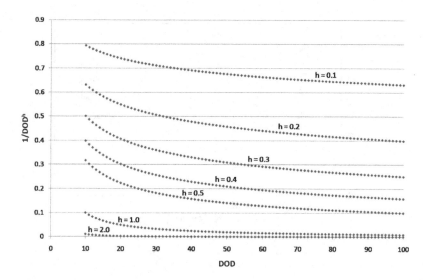

Fig. 2. $1/DOD^h$ *vs. DOD* for different h values.

Capacity vs. Number of cycles curve as also depicted in Fig. 3. From these plots, it is no simple matter to perform the battery cycle life evaluation, since the data about the number of cycles are available for a given number of DODs only (e.g., [16]) and, furthermore, sometimes they might even show an uncertainty that may range from 8 to 10 %, or even higher.

As discussed in Sect. 2, our model is meaningful if the battery under analysis only provides information in the form of two or more curves in the (capacity, number of cycles) plane, each corresponding to a different DOD.

Let us assume that there are M such curves available in a datasheet or in a measured set of data. Obviously the larger M, the more accurate the fitting process will be. Figure 3 exemplifies this scenario.

Since we need to determine two parameters from the curve(s) (h and L), and given the limited number of samples points to be considered, it is feasible to derive them from an exhaustive exploration for all C_{fade} and DOD points, as the values of h and L that minimize the maximum error with respect to the curves. However, an exploration requires a feasible range for these two parameters, which is not easy to determine because they are only weakly linked to "physical" quantities. Of the two, L is the one with some physical interpretation since it can be regarded as a correction factor of the number of cycles N. Therefore, we can assume that L ranges between 1 and a value L_{max}, determined by inspection of the datasheet. As a rule of thumb, it is usually near to the largest value of N reported in the datasheet curves. Conversely, we have no insight of possible values of h. For this reason, we implement the search as a two-phase process, as described by Algorithm 1.

The search is organized into of two main iterations over L. In the first one (Lines 1–7), for all values of C_{fade} (assumed to be discretized into P values) and

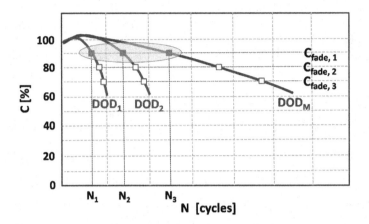

Fig. 3. Model extraction scenario.

Algorithm 1. Search for the best value of L

1: **for all** $L \in [1, L_{max}]$ **do**
2:    **for all** $C_{fade} = 1 \ldots P$ **do**
3:      **for all** $DOD = 1 \ldots M$ **do**
4:        Compute h by (8)
5:      **end for**
6:    **end for**
7: **end for**
8: $\mathcal{H} \leftarrow [h_{min}, h_{max}]$
9: $MinMaxErr \leftarrow \infty$.
10: **for all** $L = 1 \ldots L_{max}$ **do**
11:   $MaxErr \leftarrow 0$.
12:   **for all** $h \in \mathcal{H}$ **do**
13:     $TotErr \leftarrow 0, MinAvgErr \leftarrow \infty$.
14:     **for all** $C_{fade} = 1 \ldots P$ **do**
15:       **for all** $DOD = 1 \ldots M$ **do**
16:         Calculate N using (7) and compute the
               absolute error E
17:           $TotErr \leftarrow TotErr + E$
18:       **end for**
19:     **end for**
20:     $AvgErr \leftarrow TotErr/(P * M)$
21:     **if** $AvgErr < MinAvgErr$ **then**
22:       $\mathbf{H}[L] \leftarrow h$
23:       $\mathbf{Err}[L] \leftarrow AvgErr$
24:     **end if**
25:   **end for**
26: **end for**
27: $L_{opt} \leftarrow argmin(\mathbf{Err})$
28: $h_{opt} \leftarrow \mathbf{H}[L_{opt}]$

---

of the M DOD values it computes the resulting value of h using (8), which is simply a re-arrangement of (7) expressing h instead of N, and determines thus a feasible range $\mathscr{H} = [h_{min}, h_{max}]$ for h.

$$h = \frac{log(L \cdot \frac{C_{fade}}{N})}{log(DOD)} \tag{8}$$

Now that we have a feasible range for h, in the second iteration (Lines 10–26), we determine the optimal values of h and L, as follows. In the outer loop over L (Line 10), the optimal value of h is calculated first; for each value of h (using some discretization step), C_{fade} and DOD, N is computed using the model Eq. (7) (Line 16), and the error between this value and the one extracted from the datasheet is evaluated. The value of h that yields the least average error is stored as the best for a given value of L into an array **h**, together with the relative errors (array **Err**, Lines 22–23).

At the end of the iteration over L, the value of L corresponding to the smallest error is selected as single L_{opt} for the model (Lines 27–28), which is used as an index in **h** to determine h_{opt} for each C_{fade}.

4   Model Validation

The validation of the proposed model is performed after considering batteries of various chemistries produced by different manufacturers. Although the type of aging data may differ from one datasheet to another, we have collected the available information and translated it into the tabular format described in Sect. 3; using these data, we ran the search algorithm to populate the model for each battery under analysis.

4.1   VRLA Batteries

We start our evaluation from Valve Regulated Lead Acid (VRLA) batteries, which have a more evident nonlinear aging behavior with respect to many other chemistries. Moreover, datasheets for most VRLA batteries include more detailed information on aging, typically in the form of the plot of *Capacity vs. Number of cycles* (e.g., Fig. 3).

Table 1 reports the extracted manufacturer data and the resulting model parameters for two different Absorbed Glass Mat (AGM) VRLA batteries: the XTV1272 by CSB Battery and the EV12A-B by DISCOVER®. The first three columns represent the data given from the related datasheets, in both cases for three different C_{fade} points, namely 10, 20, and 40%. The last four columns report the parameters obtained by the search algorithm, the resulting number of cycles N_m from the model, and the estimation maximum absolute error. After comparing N_m against the cycle life extracted from the datasheets (i.e., N_d), the greatest errors are given for a low (i.e., 10%) C_{fade}, while they are fairly small for typical lifespan (i.e., $C_{fade} = 20\%$ or greater).

Table 1. Extracted parameters and number of cycles estimation for the CSB XTV1272 and DISCOVER EV12A-B AGM-VRLA batteries.

Battery	Datasheet			Model			
	N_d	DOD	C_{fade}	L	h	N_m	Max. error (%)
CSB XTV1272	681	30	10			597	-12.33
	305	50			1.093621	342	12.13
	151	100				160	5.96
	861	30	20	2464		770	-10.57
	374	50			1.222672	412	10.16
	186	100				177	-4.84
	1130	30	40			1021	-9.65
	459	50			1.343506	514	11.98
	231	100				203	-12.12
DISCOVER EV12A-B	1321	20	10			1512	14.46
	734	50			0.961111	627	-14.58
	348	80				399	14.66
	953	20	20	2691		2143	9.73
	885	50			1.075976	800	-9.60
	455	80				482	5.93
	2949	20	40			3017	2.31
	1071	50			1.193213	1011	-5.60
	545	80				577	5.87

Although the error is not negligible, it is worth emphasizing that very often datasheets report a possible range of the number of cycles rather than a single curve, to indicate the intrinsic uncertainty of the estimation. The spread of the values actually increases for increasing DODs. For instance, from the datasheet for the XTV1272 [16], we found that the possible variation of the cycle life (measured as the difference between the minimum or maximum value with respect to the average) might even be up to 10, 11, and 16 % for $C_{fade} = 10, 20$, and 40 %, respectively. Hence, the absolute **maximum** estimation error obtained by the proposed model (i.e., around 12, 11, and 12 %, respectively) is comparable with the maximum tolerance given by the manufacturer.

4.2   Other Battery Chemistries

Evaluation of other battery chemistries is complicated by the fact that in general only the manufacturers of VRLA batteries provide plots of *Capacity vs. Number of cycles*, for different DODs. In particular, datasheets usually report only a single curve referring to a single DOD value for lithium-based batteries. The availability of just one DOD reference, however, would yield a model with little practical use in this case, since the calibration for discharge patterns would be different from that used for characterization.

Therefore, in order to have a more meaningful assessment of the accuracy of the proposed model, we only selected those batteries whose datasheets report the

Number of cycles vs. DOD characteristic, even just for a single C_{fade} value. In any case, values of *DOD* below 10 % are not used for the derivation of the model because (i) they are not representative of typical battery usage and (ii) they are not statistically representative. It is worth noticing that the number of cycles should approach infinity as $DOD \rightarrow 0$ %; therefore, as *DOD* gets smaller it would be correct to consider a range of values rather than a precise value. Of course, all the characteristics given by the manufacturers always refer to certain operating and working conditions (e.g., charge/discharge current and temperature), which are usually different from one brand to another. In order to validate the basic proposed model, at the beginning we do not consider the differences among these conditions. However, both temperature and current rate, as stress factors in battery aging, are included in the extended model as reported in Sect. 5.

The parameters and estimation errors for the benchmark batteries are reported in Tables 2 and 3, which also report, for a more comprehensive validation, the results of the application of the existing and most meaningful analytical models [9, 11]. As (2) requires the number of cycles at DOD = 100 % as input parameter, the evaluation of that previous model was not possible for two batteries because this value is not available in their datasheets, as reported in Table 2. On the other hand, as the model proposed by [11] is useless for $DOD' = 1$ (in this case, N in (4) would be equal to zero), the analysis was re-performed by considering the maximum DOD = 80 % as reported in Table 3.

In Table 2, the largest absolute estimation error of the model occurs for a $LiFeMgPO_4$ battery, almost 20 %, while the maximum mean value is 11.35 % for the Alpha® one. However, the total average error of the maximum errors for the 10 batteries in the table is 10.66 %. The mean errors are obviously smaller, in general less than 10 %, and in one case 11.35 %.

In general, the proposed model shows robustness and accuracy for different types of electric storage devices. For the Li-ion battery by Saft Evolion the linear factor L is very high with respect to any other battery. In fact, the linear factor usually depends on the battery properties of cycling, while the range of the h parameter strictly depends on the linearity of the cycle life with respect to the DOD. The lowest h coefficient found in the model validation is 0.225627 for the Discover 22-24-700 battery, whereas the highest h is 2.000414 for the Saft Evolion.

In order to give a more comprehensive example about accuracy, Fig. 4 shows the plots obtained from all the information in the datasheet for the Lithium Manganese Dioxide Maxell ML2016 battery, and the estimation data produced by the proposed model.

Figure 5 reports the plots, normalized to the C_{fade} and parameter L, of the models for the selected batteries. The plot for the DISCOVER 22-24-6700, whose model has $h = 0.225627$, is reported separately in the upper right pane for the sake of clarity. The others are represented in a descending order of the h parameters reported in the fifth column of Table 2, i.e., the curve for the lowest value (0.995693) is at the top while the one for the highest value (2.000414) is at the bottom.

Table 2. Battery data, prediction model parameters, and estimation error of the cycle life for various batteries whose manufacturers provide the Number of cycles vs. DOD characteristic.

Producer	Code	Type	Model				[9]			
			Proposed							
			L	h	Abs. error [%]		N_1	α	Abs. error [%]	
					max	mean			max	mean
EnerSys	65-PC1750	AGM-VRLA	9083	1.393212	12.34	8.05	330	2.488793	63.03	34.49
Concorde	Sun Xtender	AGM-VRLA	4629	1.176563	15.20	8.79	354	2.644044	28.56	15.73
Sonnenschein	A600	Gel-VRLA	3874	1.020317	2.03	0.92	718	1.747624	21.12	12.84
Alpha Tech.	KL, KM, KH types	$NiCd$	31107	1.587189	18.10	11.35	463	2.412794	54.04	28.38
C&D Tech.	LI TEL 48-170 C	Li-ion	109882	1.420135	6.27	3.60	2987	2.022832	2.53	1.22
Saft	Evolion	Li-ion	1157452	2.000414	13.84	8.15	n.a	-	-	-
Seiko (SII)	MS621	$Mn\,Si\,Li-ion$	986	0.995693	0.90	0.38	202	1.712398	20.29	12.07
Maxell	ML2016	Li/MnO_2	2393	1.566125	11.28	6.49	39	2.743101	65.11	36.81
Discover	22-24-6700	$LiFePO_4$	671	0.225627	6.99	4.36	n.a	-	-	-
Valence	U-CHARGE	$LiFeMgPO_4$	153425	1.491094	19.66	9.21	2679	2.764444	19.10	12.31

Note. n.a.: not available

Table 3. Battery data, prediction model parameters, and estimation error of the cycle life for various batteries whose manufacturers provide the Number of cycles vs. DOD characteristic. The maximum DOD is 80 % for all the analyses.

Producer	Code	Type	Model							
			Proposed				[11]			
					Abs. error [%]				Abs. error [%]	
			L	h	max	mean	A	P	max	mean
EnerSys	65-PC1750	AGM-VRLA	9083	1.393212	12.34	7.49	0.00140	-0.436228	36.57	22.68
Concorde	Sun Xtender	AGM-VRLA	4629	1.176563	15.19	8.81	0.00180	-0.953029	8.37	5.44
Sonnenschein	A600	Gel-VRLA	3874	1.020317	2.03	0.82	0.00140	-1.010028	3.64	0.99
Alpha Tech.	KL, KM, KH types	$NiCd$	31107	1.587189	18.10	10.70	0.00110	-0.674032	13.67	7.29
C&D Tech.	LI TEL 48-170 C	Li-ion	109882	1.420135	6.26	4.21	0.00020	-0.864030	9.94	4.43
Saft	Evolion	Li-ion	1157452	2.000414	13.84	8.15	0.00010	-0.452045	59.81	34.84
Seiko (SII)	MS621	$Mn\, Si\, Li - ion$	986	0.995693	0.90	0.40	0.00500	-0.999028	0.99	0.42
Maxell	ML2016	Li/MnO_2	2393	1.566125	11.28	6.53	0.00500	1.228006	48.20	32.71
Discover	22-24-6700	$LiFePO_4$	671	0.225627	6.99	4.36	0.00060	-1.200934	52.83	38.52
Valence	U-CHARGE	$LiFeMgPO_4$	153425	1.491094	19.66	8.80	0.00020	-0.967028	27.47	15.69

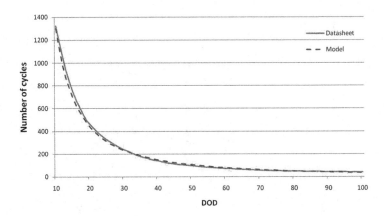

Fig. 4. Extracted N *vs.* DOD plots for the lithium manganese dioxide Maxell ML2016 battery.

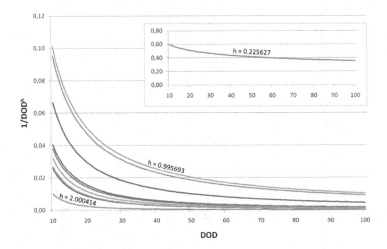

Fig. 5. $1/DOD^h$ *vs.* DOD of the generated models for the selected batteries.

At the end, the chart in Fig. 6 reports a comparison of the estimation models after applying each of them to the benchmarks. For a comprehensive report, it also includes the main results obtained for the analysis of the model by [10], whose estimation errors are too great to be reported. Furthermore, for the here proposed model, this chart considers the worst case (i.e., data reported in Table 2).

Although the previous models have two parameters (i.e., coefficients) in their expressions, one of them always strictly depends on the battery properties. In the here proposed model, both parameters L and h can be characterized, resulting in higher accuracy thanks to an additional degree of freedom in the modeling process.

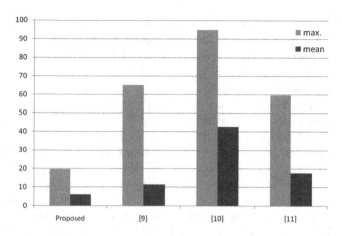

Fig. 6. Maximum and mean estimation errors given by the models for all the selected benchmarks.

5   Extension of the Basic Model

This Section provides an overview of temperature and current as stress factors that may accelerate the aging of batteries, and presents an extended version of the model reported in Sect. 3.1, in order to also include the dependency of the cycle life on these stress factors.

In this context, the total battery cycle life is the number of cycles that a battery may guarantee at different temperatures and current rates.

5.1   Impact of the Temperature on Cycle Life

The battery capacity is strongly dependent on temperature and it is not always a monotonic function. Furthermore, such a dependency changes for different battery chemistries [20].

Temperature effects on battery performance may manifest themselves in a reversible change of the total battery capacity in a single cycle, and in an irreversible capacity fading during the battery cycle life.

In the literature, an Arrhenius-type equation typically describes the relationship between battery aging due to cycle life and temperature (T) [15,22]. For fixed values of charge and discharge C-rates, this analyitical model can be written as follows [23]:

$$C_{fade} = B \cdot e^{-E_a/(R_g \cdot T_b)} \cdot A_h^z \quad (\%) \tag{9}$$

In (9), B is a constant, T_b is the battery temperature (K), while E_a and R_g are, respectively, the activation energy (J·mol^{-1}) and the universal gas constant (i.e., 8.3143 J·mol^{-1}·K^{-1}); A_h is the total ampere-hour throughput processed after a certain number of cycles (i.e., given by $N \cdot DOD' \cdot C_R$), while z is the power law factor. Regarding the latter, [15] reports that z is always near 0.5 for a graphite-LiFePO$_4$ cell, being "fairly constant at all C-rates". In addition, this work provides all the exact values of the coefficients in (9) for a certain battery cell under test. Furthermore, it should be pointed out that both input and output ampere-hour throughputs due to charge and discharge currents, respectively, contribute to capacity fading [14].

Although equations based on Arrhenius' law provide reference analytical models, nowadays batteries may have different characteristics. In fact, there are batteries for which temperature effects, in service and cycle life, do not exactly follow Arrhenius' law. For instance, the handbook for the Sonnenschein® A600 Gelled Electrolyte (GEL) VRLA battery [26] reports a better performance with respect to Arrhenius' law, from the test results, after comparing the temperature effects on both service and cycle life. Therefore, a more adaptive model that fits any characteristics concerning capacity fading should be considered when analyzing the effect of the temperature in different battery chemistries and products.

Since (9) refers to the capacity loss due to the effect of temperature in cycle life, in order to obtain a similar equation for the calendar (service) life, the term A_h in (9) must be replaced with the battery lifetime t (months) [15,22].

The Proposed Model for Temperature Effect on Aging. In order to include the temperature effect in the model, we consider a slightly different mathematical expression with respect to the model given in (7), but still with only two parameters, for extracting the temperature derating factor (TDF), as given by the following equation:

$$TDF = L_T \cdot \left(\frac{T_b}{T_{ref}} \right)^{h_T} + (1 - L_T) \tag{10}$$

In (10), T_{ref} is the temperature at which the model in (7) refers to (e.g., 25°C), while T_b is the battery temperature; L_T is an empirical constant, which appears two times in the formula, while h_t is the power factor that reflects the characteristic of the battery cycle life for different temperatures. Notice that the TDF is a non-negative value; it is in fact determined by the values of h_t and L_T using the algorithm of Sect. 3 to empirically fit the curve of N vs. T, which obviously represents a non-negative value.

5.2   Impact of the Current on Cycle Life

In various battery aging models, current is not usually considered as one of the main stress factors in cycle life (e.g., [5]). For instance, in [14] the authors claimed that the C-rate effect on aging in negligible in Li-ion cells for relatively large C-rates (in a range ±4C). This assumption cannot however be generalized for all applications and batteries. In fact, various datasheets report a different cycle life for different charge/discharge currents. For this reason, the authors in [27] proposed an extended version of Millner's aging model [5] by including both charge/discharge C-rates with their related coefficients, as extracted from the manufacturer's data for a commercial LiFePO$_4$ battery.

So, with respect to the aforementioned expression reported in (9), [15] provides a similar Arrhenius-type equation that includes the current rate (for values greater than C/2), here rewritten as follows:

$$C_{fade} = B(i_{rated}) \cdot e^{(-E_a + k_1 \cdot i_{rated})/(R_g \cdot T_b)} \cdot A_h^z \quad (\%) \qquad (11)$$

In (11), the value of the pre-exponent factor B is different for each current i_{rated} (i.e., expressed in C-rate), while E_a and z can be set to a fitted value [15], as well as the coefficient k_1.

Since charge and discharge currents usually have a different impact on aging, coefficients values in (11) are generally different when considering the charge and discharge phases.

The Proposed Model for Current Effect on Aging. The discharge current derating factor (DDF) is given by the following expression:

$$DDF = L_{i_d} \cdot \left(\frac{i_d}{i_{d_{ref}}} \right)^{h_d} + (1 - L_{i_d}) \qquad (12)$$

where $i_{d_{ref}}$ is the current (in C-rate value) to which the model in (7) refers, and $i_{d_{rated}}$ is the discharge current rate; L_{i_d} is an empirical factor, and h_d is the power factor that reflects the characteristic of the battery cycle life for different discharge rates.

Similarly, the charge current derating factor (CDF) is given by the following equation:

$$CDF = L_{i_c} \cdot \left(\frac{i_c}{i_{c_{ref}}} \right)^{h_c} + (1 - L_{i_c}) \qquad (13)$$

where $i_{c_{ref}}$ is the charge current (in C-rate value) to which the model in (7) refers to, while i_c is the discharge current; similar to the previous expression in (12), L_{i_c} and h_c are the parameters for characterizing the battery behavior for different charge rates.

Finally, the full equation for analyzing the battery cycle life as a function of DOD, T, and C-rate, is given by the following formula:

$$N(DOD, T, i) = L \cdot \frac{C_{fade}}{DOD^h} \cdot TDF \cdot DDF \cdot CDF \qquad (14)$$

In (14), both the derating factors for charge and discharge currents must be included because generally they have a different impact on battery aging and, therefore, different coefficients in their formulas.

5.3   Results

Preliminary results are obtained for the Sonnenschein A600 GEL-VRLA and Discover 22-24-6700 LiFePO$_4$ batteries, as their datasheets provide enough information for modeling their cycle life considering temperature effects. For both batteries, the analysis was conducted considering a maximum T_b equal to 50°C.

Table 4. Extracted parameters of the derating factor for the model extended to the **temperature** effect in cycle life, and consequent estimation error of the model with respect to the manufacturers' data.

Producer	Code	Type	L_T	h_T	Max. error (%)	Mean error (%)
Sonnenschein	A600	Gel-VRLA	2.99	−0.391034	9.77	3.87
Discover	22-24-6700	LiFePO$_4$	2.13	−0.840028	8.44	3.23

Table 4 reports the extracted h_T parameter for each battery, and the estimation errors of the temperature derating factor given by the model in (10) with respect to the manufacturers' data.

It is worth noticing that in both cases the maximum error is less than 10 %.

As far as concerns the current effect on battery aging, which is usually considered for high C-rates only, the model given in (12) was applied to the Discover 22-24-6700 LiFePO$_4$ battery, for which the extracted parameters L_{i_d} and h_d are, respectively, 0.98 and −0.851245. In this case, the maximum and mean estimation errors are, respectively, 2.36 % and 0.96 %. These results demonstrate the high level of accuracy that the proposed extended model may guarantee.

6   Conclusion

A compact mathematical model for estimating the number of cycles of a battery with respect to an expected capacity fade, has been proposed. The related equation describes the cycling behavior of batteries of different chemistries, and it demonstrates the possibility of obtaining a very fast and also accurate exploration of battery lifespan. The characterization of the long-term effects for a specific battery only requires two battery-specific parameters: an empirical factor L and the exponential h coefficient. Validation results show an estimation mean error generally within 10 %.

Furthermore, the basic model has been extended to include temperature and current rate effects in battery cycle life. In this scenario, various derating factors have been defined using mathematical models similar to the basic one. The mean absolute estimation errors of these models related to temperature and discharge current are, respectively, less than 4 % and about 1 %.

References

1. Benini, L., Castelli, G., Macii, A., Macii, E., Poncino, M., Scarsi, R.: Discrete-time battery models for system-level low-power design. IEEE Trans. Very Large Scale Integr. Syst. (VLSI) **9**(5), 630–640 (2001)
2. Chen, M., Rincón-Mora, G.A.: Accurate electrical battery model capable of predicting runtime and I-V performance. IEEE Trans. Energy Convers. **21**(2), 504–511 (2006)

3. Petricca, M., Shin, D., Bocca, A., Macii, A., Macii, E., Poncino, M.: An automated framework for generating variable-accuracy battery models from datasheet information. In: International Symposium on Low Power Design, pp. 365–370. IEEE/ACM (2013)
4. Peukert, W.: Über die Abhängigkeit der Kapazität von der Entladestromstärke bei Bleiakkumulatoren. Elektrotechnische Zeitschrift **20**, 20–21 (1897)
5. Millner, A.: Modeling lithium ion battery degradation in electric vehicles. In: IEEE Conference on Innovative Technologies for an Efficient and Reliable Electricity Supply, pp. 349–356 (2010)
6. Ramadass, P., Haran, B., White, R., Popov, B.N.: Mathematical modeling of the capacity fade of Li-ion cells. J. Power Sources **123**(2), 230–240 (2003). Elsevier
7. Lam, L., Bauer, P.: Practical capacity fading model for Li-ion battery cells in electric vehicles. IEEE Trans. Power Electron. **28**(12), 5910–5918 (2013)
8. Ramadesigan, V., Chen, K., Burns, N.A., Boovaragavan, V., Braatz, R.D., Subramanian, V.R.: Parameter estimation and capacity fade analysis of lithium-ion batteries using reformulated models. J. Electrochem. Soc. **158**(9), A1048–A1054 (2011). ECS
9. Seiger, H.N.: Effect of depth of discharge on cycle life of near-term batteries. In: 16th Intersociety Energy Conversion Engineering Conference, pp. 102–110. American Society of Mechanical Engineers (1981)
10. Burke, A.F.: Cycle Life Considerations for Batteries in Electric and Hybrid Vehicles. Technical paper, No. 951951. SAE (1995)
11. Thaller, L.H.: Expected cycle life vs. depth of discharge relationships of well-behaved single cells and cell strings. J. Electrochem. Soc. **130**(5), 986–990 (1983). ECS
12. Omar, N., Monem, M.A., Firouz, Y., Salminen, J., Smekens, J., Hegazy, O., Gaulous, H., Mulder, G., Van den Bossche, P., Coosemans, T., Van Mierlo, J.: Lithium iron phosphate based battery - assessment of the aging parameters and development of cycle life model. Appl. Energy **113**, 1575–1585 (2014). Elsevier
13. Symons, P.: Life estimation of lead-acid battery cells for utility energy storage. In: Fifth International Conference on Batteries for Utility Storage. Puerto Rico Electric Power Authority (1995)
14. Marano, V., Onori, S., Guezennec, Y., Rizzoni, G., Madella, N.: Lithium-ion batteries life estimation for plug-in hybrid electric vehicles. In: Vehicle Power and Propulsion Conference, pp. 536–543. IEEE (2009)
15. Wang, J., Liu, P., Hicks-Garner, J., Sherman, E., Soukiazian, S., Verbrugge, M., Tataria, H., Musser, J., Finamore, P.: Cycle-life model for graphite-LiFePO4 cells. J. Power Sources **196**(8), 3942–3948 (2011). Elsevier
16. CSB Battery Co., Ltd. http://www.csb-battery.com/upfiles/dow01404206487.pdf. Accessed 7 Jan 2016
17. Vetter, J., Novák, P., Wagner, M.R., Veit, C., Möller, K.-C., Besenhard, J.O., Winter, M., Wohlfahrt-Mehrens, M., Vogler, C., Hammouche, A.: Ageing mechanisms in lithium-ion batteries. J. Power Sources **147**(1–2), 269–281 (2005). Elsevier
18. Broussely, M., Biensan, P., Bonhomme, F., Blanchard, P., Herreyre, S., Nechev, K., Staniewicz, R.J.: Main aging mechanisms in Li ion batteries. J. Power Sources **146**(1), 90–96 (2005). Elsevier
19. Bashash, S., Moura, S.J., Fathy, H.K.: Charge trajectory optimization of plug-in hybrid electric vehicles for energy cost reduction and battery health enhancement. In: 2010 American Control Conference, pp. 5824–5831. IEEE (2010)
20. Reddy, T.B.: An introduction to secondary batteries. In: Linden, D., Reddy, T.B. (eds.) Linden's Handbook of Batteries, 4th edn. McGraw-Hill Co, New York (2011)

21. Ning, G., White, R.E., Popov, B.N.: A generalized cycle life model of rechargeable Li-ion batteries. Electrochim. Acta **51**(10), 2012–2022 (2006). Elsevier
22. Spotnitz, R.: Simulation of capacity fade in lithium-ion batteries. J. Power Sources **113**(1), 72–80 (2003). Elsevier
23. Petricca, M., Shin, D., Bocca, A., Macii, A., Macii, E., Poncino, M.: Automated generation of battery aging models from datasheets. In: 32nd IEEE International Conference on Computer Design, pp. 483–488. IEEE (2014)
24. Bocca, A., Sassone, A., Shin, D., Macii, A., Macii, E., Poncino, M.: An equation-based battery cycle life model for various battery chemistries. In: 2015 IFIP/IEEE International Conference on Very Large Scale Integration, pp. 57–62. IEEE (2015)
25. Broussely, M., Herreyre, S., Biensan, P., Kasztejna, P., Nechev, K., Staniewicz, R.J.: Aging mechanism in Li ion cells and calendar life predictions. J. Power Sources **97**, 13–21 (2001). Elsevier
26. GNB Industrial Power: Sonnenschein®: Handbook for Stationary Gel-VRLA Batteries Part 2: Installation, Commissioning and Operation, 17th edn. Exide Technologies (2012)
27. Bocca, A., Sassone, A., Macii, A., Macii, E., Poncino, M.: An aging-aware battery charge scheme for mobile devices exploiting plug-in time patterns. In: 33rd IEEE International Conference on Computer Design, pp. 407–410. IEEE (2015)

A SAR Pipeline ADC Embedding Time Interleaved DAC Sharing for Ultra-low Power Camera Front Ends

Anvesha Amaravati$^{(\boxtimes)}$, Manan Chugh, and Arijit Raychowdhury

School of Electrical and Computer Engineering, Georgia Institute of Technology, Atlanta, USA
aamaravati3@gatech.edu

Abstract. The growing need for ultra-low power cameras for sensors, surveillance and consumer applications has resulted in significant advances in compressed domain data acquisition from pixel arrays. In this journal we present a novel 64-input Successive Approximation (SAR) Pipeline analog-to-digital converter (ADC) suitable for compressed domain data acquisition in camera front-ends. The proposed architecture features a time interleaved capacitive digital-to-analog converter (DAC) shared between column parallel ADCs for area savings (2.28X); and a shared amplifier stage for power savings (60%), achieving 4X throughput as compared to traditional architectures. Simulations on a 130 nm foundry process shows that the proposed SAR Pipeline ADC draws 31 µW at 2 MS/s having a target Figure-of-Merit (FOM) of 87 fJ/conv. per step at Nyquist rate. The proposed compressive sensing front end achieves per patch energy per patch of 0.9 nJ.

1 Introduction

Mobile devices for IOT (Internet of Things) require CMOS image sensor (CIS) with low power and area [1]. Traditional CIS for wearable devices consume power more than 50 mW [2]. In a CMOS image sensor system the most power consuming blocks are: digital image processing back end & column parallel ADCs [3,4]. In most of the reported image sensors, column parallel ADCs draw 50–65% of the power of the entire image sensor signal acquisition chip [1,5]. The power consumed by column parallel ADCs is proportional to the number of measurements to be performed by the ADC. It increases with the number of pixels. For next generation IoT devices like "always on" Camera based image sensors, human machine interface systems with built in machine intelligence, low power is the key enabler.

Figure 1 shows the traditional nyquist domain signal processing. Pixel voltages are digitized using high speed column parallel ADCs. Digitized image is encoded using algorithms like discrete cosine transform (DCT), discrete wavelet transform (DWT) etc. The power budget for transmitter blocks is shown in

© IFIP International Federation for Information Processing 2016
Published by Springer International Publishing AG 2016. All Rights Reserved
Y. Shin et al. (Eds.): VLSI-SoC 2015, IFIP AICT 483, pp. 131–149, 2016.
DOI: 10.1007/978-3-319-46097-0_7

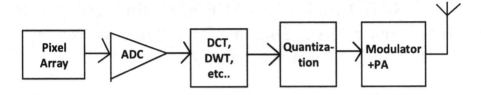

Fig. 1. Traditional nyquist signal acquisition and transmission

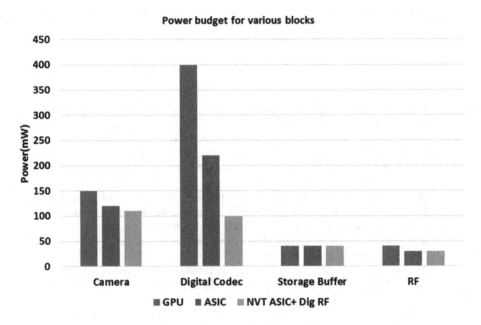

Fig. 2. Power budget for various blocks in transmitter

Fig. 2. We can observe that encoding part like DCT, DWT consumes significant amount of power followed by Analog to Digital signal acquisition etc. As the resolution of the image goes up the number of measurements per ADC goes up and hence the encoding power also increases. This places huge power constraint on acquisition device and transmitter.

Recently developed algorithms of compressive sensing (CS) promise to reduce the number of measurements with non-linear recovery at the back-end [6]. The signal processing chain for compressing sensing is shown in Fig. 3. This approach makes the encoding done at the transmitter simpler by completely eliminating power hungry blocks like DCT, DWT. If the pixel values in a camera are represented as a discrete time signal $X = [x_1 x_2 x_3 x_4 \cdots x_n]^T$, the number of measurements needed in traditional column parallel ADCs will be equal to n. Instead of n samples, CS needs only m linear measurements ($m << n$). Figure 4 shows the plot of PSNR of the recovered image w.r.to number of measurements done at receiver. We can observe that to achieve PSNR of 30 dB, 250 measurements

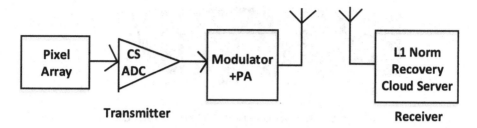

Fig. 3. Compressed domain signal acquisition & transmission

are sufficient. PSNR of 30 dB is sufficient for classifying objects [11]. Therefore the value of m can be as small as n/250. Therefore compressive sensing achieves significant reduction is encoding power & transmission bandwidth. The CS measurement matrix is given by Eq. 1.

$$Y[m] = \phi[m,n] \times X[n] = \begin{pmatrix} 0 & 1 & \cdots & 1 \\ 1 & 1 & \cdots & 0 \\ \vdots & \vdots & \ddots & \vdots \\ 1 & 0 & \cdots & 0_{m,n} \end{pmatrix} \times \begin{pmatrix} x_1 \\ x_2 \\ \vdots \\ x_n \end{pmatrix} \tag{1}$$

Here $Y[m]$ is the m-dimensional measured array, ϕ is a random binary matrix of size $m*n$ and follows the "Independent and Identically Distributed (IID)" property. X is traditionally recovered at the back-end using an optimization algorithm, like determining the L_1 norm [6].

In this paper we present a novel pipeline-SAR ADC architecture with capacitive DAC sharing with the capability of acquiring linear combinations of 64 pixel data in a single conversion cycle. This is suitable for such compressed domain data acquisition.

2   ADC Architectures for CS Image Acquisition

In prior work for obtaining compressed domain data, both analog and digital techniques have been used to perform compressive measurements from the raw data. Typically, analog implementations of compressed sensing require an analog to digital converter to improve accuracy & digital transmission [7,8]. Resistor based compressed sensing multiplexor reported in [9], suffers from static power dissipation and the number of inputs (n) is limited, making it suitable for RF receiver applications only.

To overcome some of the disadvantages of analog CS circuits, [10] has proposed compression in the digital domain after Analog to Digital Conversion. Figure 5 (a) shows the technique proposed in [10]. The entire analog signal is converted into the digital domain by high-speed ADCs and the CS encoder does compression in the digital domain. This is primarily suited for low bandwidth application like bio-medical signal processing. However, for CIS of a typical

Original Image   M=250   M=150   M=100
(256x256)

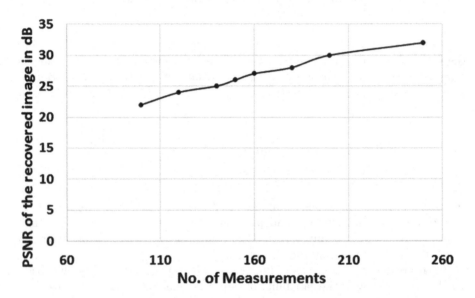

Fig. 4. Recovered image using L1 norm

256 ∗ 256 size, ADCs would need to acquire all the samples and then convert to the digital domain. The number of measurements by the ADC will not be reduced and it defeats the purpose of compressed domain data acquisition. Therefore ADC power will remain the same for image acquisition. Further, the size of digital CS encoder grows exponentially with the number of inputs. CS encoders will further add significant power along with the ADC making it infeasible for "always on" imaging front-end applications (Fig. 6).

To overcome the limitations of data acquisition followed by compressed domain measurements, Oike et.al, has proposed a CS camera through simultaneous averaging and quantization of pixels using a $\Sigma - \Delta$ ADC [5]. Figure 5 (b) shows the schematic of the resetting $\Sigma - \Delta$ ADC used for such linear measurements. Pixel values are multiplied with random numbers (from the ϕ matrix) sequentially and passed to the input of the $\Sigma - \Delta$ ADC. This approach requires m measurements; however it requires n conversion cycles for one measurement. This architecture requires a 16 ∗ 16 block for linear measurement. For each

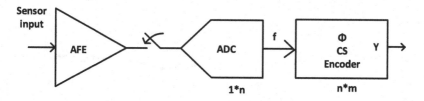

Fig. 5. CS encoder on acquired samples of ADC [10]

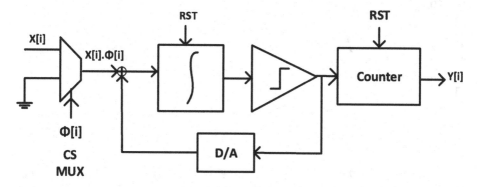

Fig. 6. Simultaneous compression and quantization within [5]

measurement of the block, the resetting $\Sigma - \Delta$ ADC needs 256 clock cycles. For m measurements $\Sigma - \Delta$ ADC needs $n * 256$ clock cycles. During this conversion period, all the high gain amplifiers will remain on and consume power. Hence, for lowering the total power dissipation, faster conversion with the opportunity for power gating once the conversion is complete, will be critical.

Once compressed domain data is acquired, the image is often used for online classification to detect potential trigger signals. For such in-situ classification [16] and trigger identification, 8 bits of inputs are sufficient. We have plotted classification accuracy vs. Bit resolution for MNIST data base in Fig. 7. We can observe that recognition accuracy becomes constant after more than 6 bits of resolution. Further, it has been shown that for most of the machine learning applications moderate resolution (6–8 bits) is sufficient [1,11]. Figure 8 shows the Energy per conversion with respect to the Signal to Noise and Distortion ratio (SNDR) for state of the art SAR, Pipeline and $\Sigma - \Delta$ ADCs. SNDR is related to effective number of bits ($ENOB = SNDR - 1.76/6$) [19]. From this plot we can observe that SAR ADC has best FOM (order of 10–1000 fJ/conv) for moderate resolution (6–8 bits). Pipeline ADCs also have competitive FOM for moderate and high speed applications. Since most of the image sensors speed varies from 1 MS/s to 10 MS/s and we are interested in 8 b of resolution for in-situ image processing applications, we propose a SAR-Pipeline ADC which achieves ultra-low power and high area efficiency.

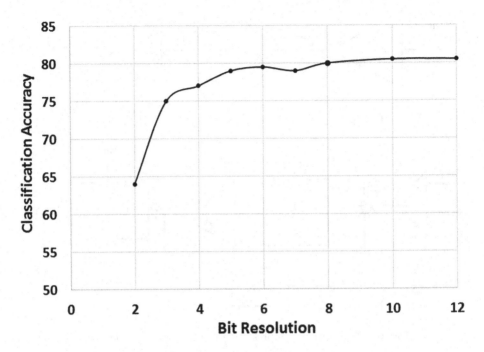

Fig. 7. Classification accuracy vs. Bit resolution for MNIST database

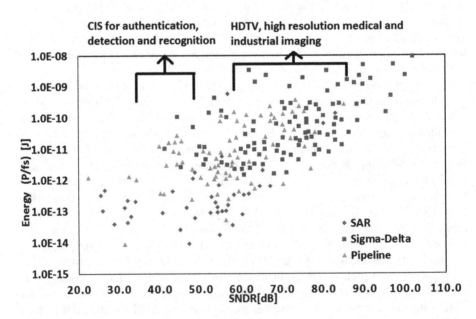

Fig. 8. Energy vs. SNDR for state of the art reported SAR, Pipeline and $\Sigma - \Delta$ ADCs [15]

3   SAR-Pipeline ADC Architecture for CS Measurements

For most of the low power applications SAR ADCs are used since they consume ultra-low energy per conversion (Fig. 8). However, for portable image front-end applications resolution more than 4–5 bits SAR ADC occupies huge area since the MSB capacitance grows as 2^N. Since there will be many column parallel ADCs each will have capacitance of 2^N. To alleviate this problem two-stage SAR Pipeline ADCs are proposed [20,21]. SAR-Pipeline uses two stage SAR-ADC and an amplifier which is used for amplifying residue generated by stage 1 SAR ADC (Fig. 9). Both the SAR ADC stages operates in parallel and each stage has to resolve lesser number of bits (lesser DAC settling time & capacitance (hence lesser area) as compared to traditional SAR). Therefore, SAR-Pipeline ADCs can operate at much higher speeds with high area efficiency [21]. One of the inherent advantage of SAR-Pipeline is residue voltage of Stage-1 SAR ADC is generated within its DAC after conversion phase. Hence this avoids extra DAC and clock phase to generate residue of Stage-1 unlike in traditional flash based Pipelined ADCs [21].

Figure 9 illustrates the proposed ADC architecture. The design operates on a block size of 16 * 16 (256 elements in pixel array). We propose SAR-Pipeline ADC consisting of 64-inputs Stage-1 SAR ADC resolving 4 bits (with 1 bit redundancy) and Stage-2 SAR ADC resolving 5 bits. We propose time-interleaved DAC sharing for Stage-1 SAR ADC which provides a linear measurement of 64-inputs in a single conversion cycle. We also share the amplifier (used for residue amplification) between 2 neighboring column parallel ADCs to save power. 64 inputs are simultaneously averaged and quantized using the SAR-Pipeline ADC. Post-conversion, 4 consecutive samples are averaged using a 10 bit accumulator and shift register. This allows us to average 256 samples in 4 ADC conversion cycles. For m random measurements ADC takes $m * 4$ conversion cycles.

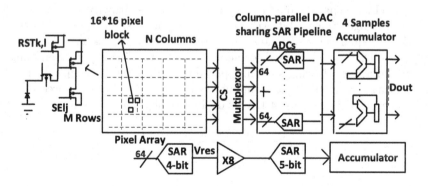

Fig. 9. Proposed CS front-end architecture for CIS

Figure 10 shows a previously reported multi-input SAR ADC used for compressed sensing (with 8 bit resolution). It uses charge sharing. The MSB capacitor is equally divided among the inputs. Because of charge sharing the inputs will get averaged after the sampling cycle. [17] demonstrates a 4 input CS SAR ADC for wireless applications. This technique requires $(2^8 + 2^4 = 272C)$ number of capacitors for an 8 bit ADC and measures 256 inputs (C is the unit capacitor). One of the main limitations of the proposed SAR ADC architecture for portable imaging application is the area occupied by the sampling capacitors [18]. Dividing the MSB capacitors to accommodate 256 inputs requires 256 switches. For portable applications limited supply $\approx$1–1.3 V provides high R_{ON}. This provides us the time constant (τ_{conv}) for conversion (min. sized capacitor of 50 fF) of $\approx$220 ns (DAC settling time). This allows a maximum sampling frequency of 730 KHz. Hence, for high speed cameras (with 30 frames/s) the proposed ADC architecture will not be able to meet latency requirements. Further, [17] uses calibration for capacitor mismatch, a requirement for more than 6-bits of resolution.

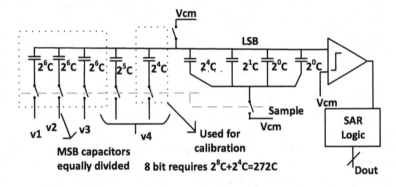

Fig. 10. Reported multi-input SAR-ADC [17]

Figure 11 is the proposed SAR-Pipeline with DAC sharing. We use 4 bit ADC as the first stage. Since 4-bit ADC has 16 C capacitors, all the capacitors are divided into equal value of C and 16 inputs are applied. We have 3 instances of the same DAC which is used for accessing additional 48 inputs. Sampling is done in two phases. During sampling phase (S1) all 4 DAC's sample 16 inputs each. During second phase of sampling charge is redistributed between them. The averaged voltages across 4 DAC's during S1 phase given by Eq. 2.

$$V_{dac1} = \frac{(v_1 + v_2 + ... + v_{16})}{16}$$

$$\vdots \tag{2}$$

$$V_{dac4} = \frac{(v_{48} + v_2 + ... + v_{64})}{16}$$

During the second sampling phase S2, averaging of V_{dac1} to V_{dac4} takes place. Therefore, the final voltage across DAC is given by Eq. 3.

$$V_{dacf} = \frac{(V_{dac1} + ... + V_{dac4})}{4}$$

$$V_{dacf} = \frac{(v_1 + ... + v_{64})}{64} \tag{3}$$

$$V_{dacf} = \frac{(X[0].\phi[0] + ... + X[63].\phi[63])}{64}$$

We can observe form Eq. 3 that the final accumulated output represents the dot-product of the input pixel vector X with the sampling matrix, ϕ. ϕ can be random or programmed so that both random as well as structured compressed measurements can be obtained. As soon as S2 is done 3 DAC's are shared with neighboring column parallel ADC. Once the conversion in 4-bit SAR ADC is complete, we amplify the residue by 4× and pass it to a 5-bit fine ADC to resolve the LSBs. Ideally a gain of 16 is required for residue amplification. We use 1-bit digital redundancy in Stage 1 and half reference scaling for Stage 2 to reduce the gain requirement which helps to reduce the power in the high-gain op-amp [20].

Since all the capacitors we use are identical and of value C, calibration is not required (more details in section III). As 3 DAC's are shared with 4 ADC's, we need an additional capacitance of 12 C. With 12 C extra capacitance we can acquire linear measurements of 64 inputs in each conversion cycle. This DAC shared method significantly improves area efficiency and enables simultaneous acquisition of multiple inputs. In this architecture, the conversion time-constant (τ_{conv}) is determined by the 4-bit ADC settling time even tough we are sampling 64-inputs. This makes the architecture suitable for high speed sensing with large number of inputs.

Figure 12 shows how the sampling schemes are time-interleaved for the entire column parallel ADC architecture. Conversion cycle for ADC is 8 clock cycles. During this period we share 3 DACs with 3 of the neighboring ADCs. S3 to S8 are sampling phases of ADC2 to ADC4. S3 to S8 phase operates during conversion period of ADC1. Pipelining facilitates overlapping of Stage-1 and Stage-2 conversion phases. 1-bit redundancy is added in the first stage to accommodate capacitor mismatch and offsets of the comparator, amplifiers [20]. We also share residue amplifier between two neighboring ADCs to reduce the total power [21]. Accumulator (10 bit) used to average 4 consecutive ADC output samples. The accumulator is reset after every 4 sampling cycles (F_s). The sampler operating at quarter sampling rate is used to capture the averaged output. The averaged output contains random measurement of 256 inputs. Figure 12 also shows the control logic used for proposed CS front-end ADC architecture. Global reset (RST) is used generate S1, S2 and conversion phase for ADC1. S2 phase of ADC1 is used to trigger sampling phase for neighboring column parallel ADC. This process is continued for all 4 ADCs. Falling edge of S2 phase triggers the conversion phase of individual ADCs (Fig. 13).

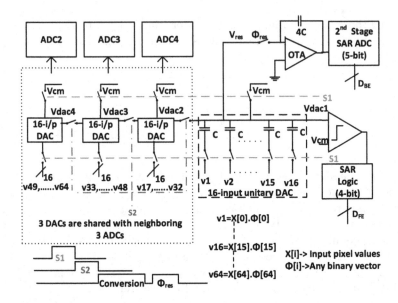

Fig. 11. Proposed multi-input DAC sharing SAR ADC

4   Design Components

In this section, the design details of the first and the second state of the ADC are discussed.

4.1   Stage 1 ADC and Residue Amplification

Figure 14 shows the Stage 1 of the proposed SAR-Pipeline ADC. 64 inputs are acquired from S1 and S2. Residue is fed into an amplifier with gain of 4. Stage 1 of the ADC has 4 bit resolution with 1 bit digital redundancy. 1 bit redundancy is used to accommodate the residual offset of the comparator, op-amp and capacitor mismatch errors.

The Op-amp open loop gain (A_{OL}), unity gain frequency (f_u) and swing ($Vp - p$) target based on the inter-stage gain is given in Table 1. The required values are derived as per gain error, gain bandwidth (GBW) requirement of the OTA to be within $1/2$ LSB of the ADC error [21]. The worst case values across process corners is mentioned in the Simulated values of the Table. We can observe that, simulated values across process corners for gain, bandwidth are by a factor of two larger than required values. Figure 15 shows the telescopic cascode OTA used as interstate amplifier. It is well suited for two stage pipeline SAR since the swing requirement is low and it has high gain bandwidth efficiency.

We use pre-amplifier with output offset compensation to limit the offset of Stage 1 SAR ADC. The residual offset ($V_{os,res}$) is given by Eq. 4.

$$V_{os,res} = \frac{V_{os,pre-amp}}{A_p} + \frac{V_{os,latch}}{A_p} \tag{4}$$

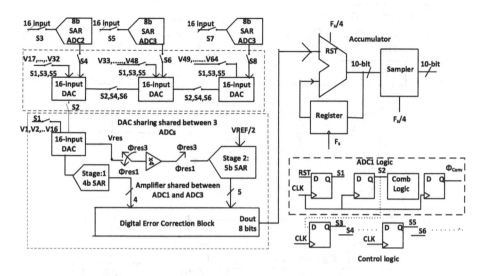

Fig. 12. Proposed SAR ADC with time-interleaved DAC sharing

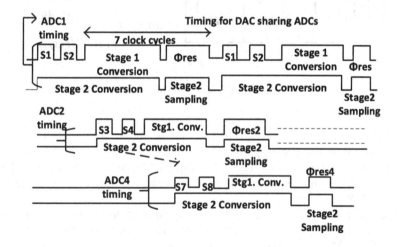

Fig. 13. Timing for proposed SAR Pipeline Architecture

Table 1. Design requirement for amplifier and 2nd Stage offset

Inter-stage gain	Op-amp			2nd stage SAR
	A_{OL}	f_u	$Vp - p$	Offset
Required	42 dB	42 MHz	250 mV	16.125 mV
Simulated values	50 dB	80 MHz	300 mV	8 mV

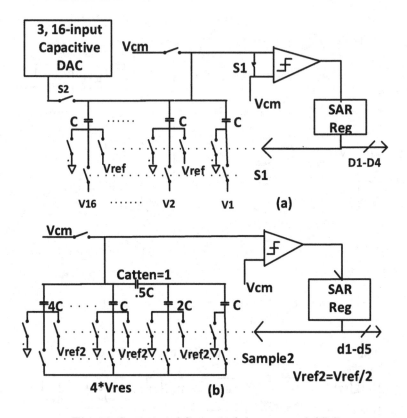

Fig. 14. Stage 1 and Stage 2 of the proposed ADC

where $V_{os,pre-amp}$ and $V_{os,latch}$ are the pre-amplifier offset and latch offset respectively. A_p is the pre-amplifier gain. The 3σ $V_{os,pre-amp}$ and $V_{os,latch}$ are 5 mV and 30 mV respectively. The gain amplifier features a cross coupled load which provides a high gain of 15. The residual offset is 2.33 mV which is 0.25 LSB of the sub-ADC.

We use telescopic cascoded OTA in the proposed design for residue amplification. Telescopic cascoded OTA has high power efficiency for a given gain bandwidth (GBW) [19]. Because of half gain and half reference implementation of the ADC, the open loop gain of the OTA is reduced. The OTA achieves a swing of 300 mV$_{p-p}$.

4.2 Stage 2 ADC

Figure 16 shows the Stage 2 of the proposed SAR-Pipeline ADC. We use a split capacitor architecture to reduce the area and power for the second ADC. Since the non-linearity of this ADC will get divided by the gain of the amplifier, it can be neglected. For the comparator in stage 2 of the proposed ADC, a pre-amplifier with gain of 3 is used since the offset requirement from it is 15 mV. Hence, Stage 2

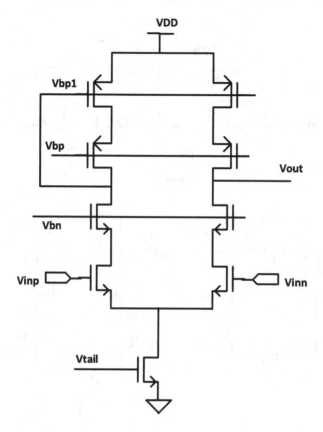

Fig. 15. Telescopic cascode OTA used as inter-stage amplifier

of the proposed ADC doesn't require the output offset compensation. The total capacitance from the second ADC is 11 C.

5   Analysis of Capacitor Mismatch

Systematic variations has no effect of capacitor matching since all the capacitance in Stage 1 SAR ADC are equal to C. The capacitance mismatch standard deviation for metal-insulator-metal (MiM) is given by Eq. 5.

$$\sigma_{\Delta C/C} = \frac{A_{\Delta C/C}}{\sqrt{WL}} \tag{5}$$

where $A_{\Delta C/C}$ is process constant which is $1\,\%.\mu$m for $0.13\,\mu$m CMOS process [22]. W &/ L are width and length of the capacitor. The minimum size allowed in $0.13\,\mu$m is $5\,\mu$m $*5\,\mu$m. With minimum sized capacitor $\sigma_{\Delta C/C}$ obtained will be 0.002.

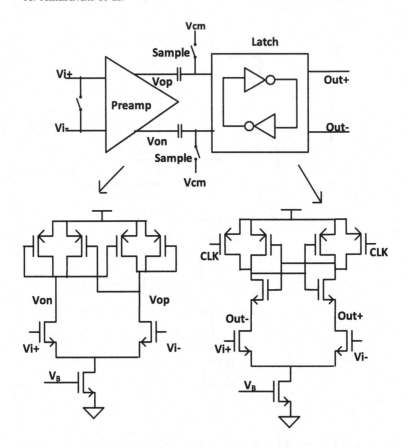

Fig. 16. 6-bit split cap SAR-ADC for Stage 2

As per [23] maximum allowable capacitor mismatch for a resolution of n is given by Eq. 6.

$$\frac{\Delta C}{C_{max}} = \frac{2^n}{2^{2n} - 2^n + 1} \tag{6}$$

For n = 9, $\Delta C/C_{max}$ reaches close to 0.002. This shows the residue generated by first ADC will fall within the range of 1/8 LSB of error. Hence the proposed architecture is robust towards capacitor mismatch.

6   Simulation Results

Performance of the proposed SAR-Pipelined ADC is verified through design and simulations in the 0.13 μm Mixed-Mode CMOS.

Figure 17 shows the normalized output frequency spectrum of the proposed ADC for input frequency (F_{in}) of 248.34 KHz at sampling rate (Fs) of 1 MSPS. A 1024-point FFT shows SNDR of 49.5 dB which is equivalent to an ENOB of 7.9.

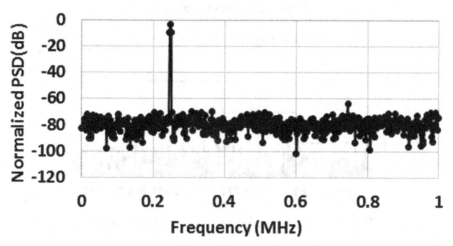

Fig. 17. Frequency spectrum of the proposed ADC (Fin = 248.34 KHz & Fs = 2 MHz)

Figure 18 shows the 64 inputs applied to ADC at each sampling cycle. Each 64 inputs corresponds to CS multiplexor output (Product of input vector with random number). Figure 19 shows the ADC and accumulator outputs at each conversion cycles. For a particular case study, as shown in the figure, an ideal averaging without quantization results in a output of 270.11 mV. The proposed ADC after accumulated 4 samples each provides an output of 269.53 mV which is less than 1 LSB of error.

Figure 20 shows the SNDR of the proposed ADC from input frequency range of 0.2 MHz to to 0.98 MHz. The ENOB at Nyquist frequency is 7.56. This ENOB achieves Walden FOM [19] of 85 fJ/conv. step.

Figure 21 shows the DNL and INL of the proposed ADC across 256 digital codes. The worst case DNL is within 0.4 LSB. INL is within 1 LSB across all digital codes.

7 Power Budget & Energy Efficiency

The power budget for the proposed ADC is given in Table 2. The power number is w.r.to patch size of 16 ∗ 16 and a compression ration (CR) of 16. Even though the total power consumed from the supply is 50 μW, since the amplifier is shared between two ADC, the power for individual ADC's is 31 μW. The number of conversion cycles required for 16 ∗ 16 patch size with compression ratio of 16 is $(16 * 16 * 16)/64 = 64$.

The energy per patch is given by Eq. 7.

$$Energy_{Patch} = \frac{P * N_c}{F_s} \qquad (7)$$

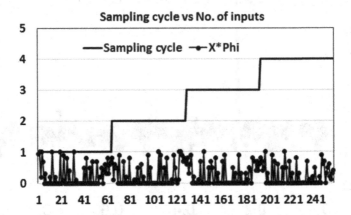

Fig. 18. 64 inputs for ADC for every 1 sampling cycle

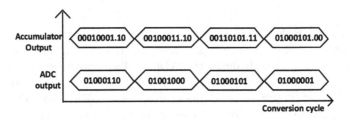

Fig. 19. Output of ADC and accumulator for 4 conversion cycle

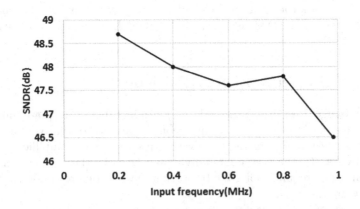

Fig. 20. Simulation result of SNDR vs. Input frequency at $Fs = 2\,MHz$

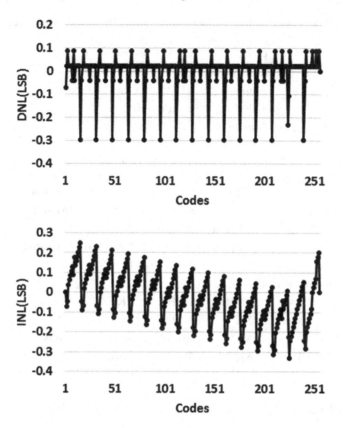

Fig. 21. DNL and INL of the proposed ADC across 256 codes

where, P is the power drawn by ADC for each conversion, N_c is the number of conversion cycles & F_s is the sampling frequency. The energy per patch for the proposed design is 0.9 nJ.

8   Comparison with Reported Works

Table 3 shows the comparison of the proposed design with state of the art CS architecture. Proposed design is scalable and can handle a large number of inputs at the same time. Due to parallelism achieved by sharing DACs between columns parallel ADCs high energy efficiency per patch is achieved.

Table 2. Power and capacitance contribution from individual blocks

Block	Power/	Capacitance
SAR Stage 1	7 μW	28 C
Amplifier	41 μW	4 C
SAR Stage 1	2.5 μW	10 C
Accumulator	0.5 μW	Nil

Table 3. Comparison with reported works

	Oike [5]	Guo [17]	Chen [10]	This work
ADC type	$\Sigma - \Delta$	SAR	SAR	SAR-Pipeline
Technology	0.15 μm	0.13 μm	0.09 μm	0.13 μm
Design	Measured	Simulated	Measured	Simulated
No. of inputs	1	4	1	64
Sampling cycles	256	1	256	4
F_s	1 MHz	1 MHz	2 KHz	2 MHz
Capacitance	NA	272 C	256 C	40 C
Power	NA	50 μW	5 μW	31 μW
Energy/Patch	NA	51 nJ	640 nJ	0.9 nJ

9   Conclusion

Multiple techniques are proposed to achieve high throughput in column parallel ADCs used for image sensors. Time interleaved sharing DAC technique reduces the number of measurement required by a factor of 4. Sharing the amplifier between neighboring column parallel ADCs reduces the power by 64 %. The proposed architecture can be used for wearable devices with ultra-low power requirements. Our design and simulation results show 87 fJ/conv. step with an average power of 31 μW.

References

1. Choi, J., Sin, J., Kang, D., Park, D.: A 45.5 μW 15 fps always-on CMOS image sensor for mobile and wearable devices. In: IEEE International Solid-State Circuits Conference (ISSCC) Digest of Technical Papers, pp. 114–117 (2015)
2. Deguchi, J., Tachibana, F., Morimoto, M., Chiba, M.: A 187.5 Vrms-read-noise 51 mW 1.4 Mpixel CMOS image sensor with PMOSCAP column CDS and 10 b self-differential offset-cancelled pipeline SAR-ADC. In: IEEE ISSCC 2012, pp. 494–496 (2012)
3. Park, J., Aoyama, S., Watanabe, T., Isobe, K., Kawahito, S.: A high-speed low-noise CMOS image sensor with 13-b column-parallel single-ended cyclic ADCs. IEEE Trans. Electron Devices **56**(11), 2414–2422 (2009)

4. Watabe, T., Kitamura, K., Sawamoto, T., Kosugi, T., Akahori, T., Iida, T., Isobe, K., Watanabe, T.: A 33 Mpixel 120 fps CMOS image sensor using 12 b column-parallel pipelined cyclic ADCs. In: IEEE ISSCC Digest of Technical Paper, pp. 388–389 (2012)
5. Oike, Y., Gamal, A.: CMOS image sensor with per-column $\Sigma\Delta$ ADC and programmable compressed sensing. IEEE J. Solid State Circuits **48**(1), 318–328 (2013)
6. Donoho, D.: Compressed sensing. IEEE Trans. Inf. Theory **52**(4), 1289–1306 (2006)
7. Gruev, V., Cummings, R.: Implementation of steerable spatiotemporal image filters on the focal plane. IEEE TCAS-II **49**(4), 233–244 (2002)
8. Robucci, R., et al.: Compressive sensing on a CMOS separable-transform image sensor. Proc. IEEE **98**(6), 1089–1101 (2010)
9. Slavinsky, J.P., et al.: The Compressive multiplexer for multi-channel compressive sensing. In: Proceedings of IEEE ICASSP, pp. 3980–3983 (2011)
10. Chen, F., Chandrakasan, A., Stojanovic, V.M.: Design and analysis of a hardware-efficient compressed sensing architecture for data compression in wireless sensors. IEEE JSSC **47**(3), 744–756 (2012)
11. Lee, E., Udell, M., Wong, S.: Factorization for Analog-to-Digital Matrix Multiplication, Report. Standford University (2013)
12. Chae, Y., et al.: A 2.1 Mpixel 120 frame/s CMOS image sensor with column-parallel $\Sigma\Delta$ ADC architecture. IEEE JSSC **46**(1), 236–247 (2011)
13. Lee, E., H.: A 2.5 GHz **7**.7 TOPS/W switched-capacitor matrix multiplier with co-designed local memory in 40 nm. In: IEEE ISSCC, pp. 418–420 (2016)
14. Toyama, T: A 17.7 Mpixel 120 fps CMOS image sensor with 34.8 Gb/s readout. In: IEEE International Solid-State Circuits Conference (ISSCC), pp. 420–422 (2011)
15. Murmann, B.: ADC Performance Survey 1997–2015. http://web.stanford.edu/murmann/adcsurvey.html
16. Zhang, J., Wang, Z., Verma, N.: A matrix-multiplying ADC implementing a machine-learning classifier directly with data conversion. In: ISSCC, pp. 1–3 (2015)
17. Guo, W.: A single SAR ADC converting multi-channel sparse signals. In: Proceedings of IEEE ISCAS 2013 (2013)
18. Chen, D.: A 64 fJ/step 9-bit SAR ADC array with forward error correction and mixed-signal CDS for CMOS image sensors. IEEE TCASI **61**(1), 3085–3093 (2014)
19. Gustavsson, M., Wikner, J., Tan, N.: CMOS Data Converters for Communication. Kluwer Academic Publishers, Norwell (2000)
20. Lee, C., Flynn, M.: A SAR-assisted two-stage pipeline ADC. IEEE JSSC **46**(4), 859–869 (2011)
21. Zhu, Y.: A 50-fJ 10-b 160-MS/s pipelined-SAR ADC decoupled flip-around MDAC and self-embedded offset cancellation. IEEE JSSC **47**(11), 2614–2626 (2012)
22. Diaz, C., Tang, D., Sun, J.: CMOS technology for MS/RF SoC. IEEE Trans. Electron Devices, 81–84 (2003)
23. Lin, Z.: Modeling of capacitor array mismatch effect in embedded CMOS CR SAR ADC. In: Proceedings of 6th International Conference on ASICs, pp. 979–982 (2005)

Electromagnetic Transmission of Intellectual Property Data to Protect FPGA Designs

Lilian Bossuet[1(✉)], Pierre Bayon[2], and Viktor Fischer[1]

[1] Laboratoire Hubert Curien, CNRS UMR 5516,
Université Jean Monnet, 42000 Saint-Etienne, France
{lilian.bossuet,fischer}@univ-st-etienne.fr
[2] Brightsight, Delft 2628, The Netherlands
bayon@brightsight.com

Abstract. Over the past 10 years, the designers of intellectual properties (IP) have faced increasing threats including cloning, counterfeiting, and reverse-engineering. This is now a critical issue for the microelectronics industry. The design of a secure, efficient, lightweight protection scheme for design data is a serious challenge for the hardware security community. In this context, this chapter presents two ultra-lightweight transmitters using side channel leakage based on electromagnetic emanation to send embedded IP identity discreetly and quickly.

1 Introduction

The microelectronics industry is faced with increased costs of production of integrated circuits (ICs). This increase is due to the costly technology refinement and the increasing complexity of systems (e.g. the transition from 32 nm to 28 nm technology has been accompanied by a 40 % increase in the manufacturing costs of wafers 300 mm in diameter and by a 30 % increase in the manufacturing costs of 450 mm wafers). For several years, this led to a sharp increase in the number of companies that do not have the means to produce IC (fabless companies) and to the relocation of production. ICs manufactured today are produced with a high added value in a highly competitive industry. In addition, the time-to-market is increasingly tight. This has made expensive devices the target of counterfeiting, cloning, illegal copy, theft and malicious hardware insertion (such as hardware Trojans) [1, 2].

1.1 The Threat Model of IC and IP

The counterfeiting of ICs has become a major problem in recent years [3]. For example, the number of counterfeit electronic circuits seized by U.S. Customs between 2001 and 2011 has been multiplied by around 700 [4]. Between 2007 and 2010, U.S. Customs confiscated 5.6 million counterfeit electronic products [5]. Overall, counterfeiting is estimated to account for about 7 % of the semiconductor market [6], which represents a loss of around US$ 22 billion in 2014 for the lawful industry.

Figure 1 is a simplified diagram of the life cycle of an IC from its design by a fabless designer to its recycling. This cycle includes many threats to the designer's intellectual property: netlist theft, mask theft, chip over-production (overbuilding), theft

© IFIP International Federation for Information Processing 2016
Published by Springer International Publishing AG 2016. All Rights Reserved
Y. Shin et al. (Eds.): VLSI-SoC 2015, IFIP AICT 483, pp. 150–169, 2016.
DOI: 10.1007/978-3-319-46097-0_8

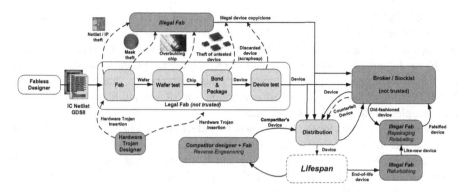

Fig. 1. Simplified life cycle of an IC from a fabless designer to device recycling and the associated threat model.

of the untested device, discarded device, reverse engineering, device counterfeiting, cloning, relabeled-repackaged-falsified "like new device", and hardware Trojan insertion.

The threat model in Fig. 1 focuses on IC manufacturing and does not include the specific case of IP design and licensing. Indeed, for digital circuit design the re-use of IP is more and more important due to prohibitive cost of ASIC design, but the IP business suffers from a lack a security due to the intrinsic form of IPs sales and exchanges. Figure 2 presents a dedicated threat model focused on an IP life cycle. Many dedicated threats target the IP life cycle and result to revenues losses for the IP designers [1, 2]. The IP threat model includes illegal re-use, illegal sales, cloning (illegal copy) of the IP. The extent of threats targeting IPs is linked to the type of IP: soft IPs (typically hardware description language files), firm IPs (*synthesized* netlist), and hard IPs (FPGA bitstream or physical layout).

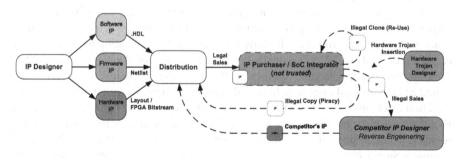

Fig. 2. Simplified life cycle of an IP with its threat model

We propose a way to counter theft, illegal copy, cloning and counterfeiting of ICs and IPs by designing a salutary hardware (*salware*) [7]. The term salware is the opposite of malware (malicious hardware). While salware can use the same techniques, strategy and means as malware [7], salware uses an embedded piece of hardware that is barely detectable, hard to circumvent, and is inserted in an IC or an IP to provide intellectual property information and/or to remotely activate the circuit or IP after its manufacture and during its use. IP watermarking, physical unclonable function (PUF) for IC authentication, remote activation, logic encryption, finite state machine (FSM) encryption, memory encryption, bus encryption, hardware metering, VHDL/Verilog obfuscation, bitstream encryption (for SRAM and Flash based FPGAs), are examples of the well-known salwares.

One of the solutions for the IP designers to protect their intellectual property is to be able to detect the presence of a copy of an IP embedded in a digital device by using IP identification. Works on IP watermarking and IP fingerprinting try to provide the IP identification service. But, most of the time the published solutions are not practical mainly because of the complexity of the watermarking/fingerprinting verification scheme [8, 9]. Efficient IP identification scheme needs to be contactless, rapid and ultra-lightweight. Up to now, these three characteristics are not available in the state-of-the-art. To meet these requirements, in this chapter we propose an ultra-lightweight binary frequency shift keying (BFSK) transmitter to forward IP identity (that could be generated for example by a feedback shift-register or a physical unclonable function [10]) discreetly using an electromagnetic channel. Such circuit is usually called "spy circuitry". Using the electromagnetic channel, it is possible to contactless check the presence of an IP inside a digital device. A preliminary version of this work was presented during the conference VLSI-SOC 2015 [11].

1.2   Salware vs. Malware

In the area of security, the techniques used to attack and to defend have always been similar and the means designed for attacks can sometimes be used for protection. Our strategy in investigating the means of attack and malicious hardware is to develop new efficient salware.

Small, barely detectable hardware Trojans can disable part of a device or allow information leakage without degrading system performance [12, 13]. The same characteristics are required to design efficient salware, which is our objective. Embedding a Trojan inside an IP to protect the IP during its time-limited evaluation by the client was recently proposed in [14]. This work modifies the FSM of the IP with the aim of disrupting its normal behavior. In this way, the IP vendor can define the "expiry date" of the FSM control and disable it. In fact, this application uses a Trojan like time-based activation mechanism [12, 13]. Other activation mechanisms can be used to disrupt the IP in the case of illegal use such as an expired hardware license or an illegal copy. For example, a Trojan-like salware can use a PUF response to conditionally block an IP execution. Such a physical-condition-based activation makes it possible to link an IP to the hardware (hardware-linked license).

Another well-known threat in cryptographic engineering is side channel attacks [15, 16]. Most of the dynamic characteristics of both hardware and software implementations of cryptographic primitives can be used for side channel analysis: computation time, power consumption, electromagnetic radiation, optical radiation, even the sound produced during computation. However, the techniques used for side channel analysis can be used to implement a salware block: e.g. for reading intellectual property data from the device or for device authentication (watermark checking). Some published works propose spy circuitry using side channels to identify the embedded intellectual property. For example in [17], the thermal channel representing a contactless communication was used to transfer information from an embedded tag to a remote receiver. However the embedded thermal tag used in this commercial solution requires a relatively large area (255 Spartan-3 slices). In [18], the authors propose using two shift registers to generate a recognizable signature-dependent power consumption pattern to reveal the IP signature. Power consumption was also used in [19] to communicate the IP watermark data using classical differential power analysis (DPA [15]). To reinforce such work, the authors of [20] propose using the power supply signal of an IP as a physical hash function for fingerprinting.

As we mentioned above, hardware Trojans can be designed to change the operation of the infected device, but also to silently leak information. Hardware Trojans can use side channels to forward secret information such as a symmetric cipher key [21] from cryptographic hardware implementation [22, 23], even when secure key management is used [24]. Hardware Trojan is also used to cause or amplify side-channel leakage of cryptographic hardware [25]. Note that using side channels to detect a hardware Trojan has also been the subject of several studies [26–28].

However, designing salware with a Trojan-like hardware could present a new opportunity to protect IC and IP. In this paper, we propose a, Trojan-like, IC/IP information provider that is discreet, contactless, ultra-lightweight, and with a high bitrate. It uses an electromagnetic side channel to transmit useful information.

Except [17], all the related works use power consumption as a communication channel which is not contactless. Unlike the proposed solution, all the related works are not lightweight and rapid as the Sect. 5 of this chapter will show.

2 EM Communication of IP Data

2.1 Principle

Previous works on the electromagnetic attacks targeting true random number generators (TRNGs) showed that electromagnetic radiation can be used very efficiently for both active (fault injection [29]) and passive (side channel analysis [30]) attacks. Compared to power analysis, the attacker measuring the near-field electromagnetic emissions can obtain additional partial information about the device, since, unlike measurement of power consumption, electromagnetic radiation can be measured locally. One of the main advantages of this side channel is that it is impossible to hide the leak concerning electromagnetic radiation by using a global countermeasure. Moreover the electromagnetic test bench is not expensive (less than US$ 10K without an oscilloscope, which is the most expensive component). Last but not least, a spectral analysis of the electromagnetic radiation provides information on the oscillating structure such as a ring-oscillator [30]. For all these reasons, we use the electromagnetic channel for our IC/IP identification scheme. To this end, we designed an ultra-lightweight BFSK transmitter.

As mentioned above, salware and malware can be based on similar principles. The same is true for the proposed BFSK principle, which can be used to design both salware (i.e. IP identity transmitter) and malware (i.e. stolen data transmitter driven by a hardware Trojan), as illustrated in Fig. 3. There are two differences between using the BFSK as salware or malware. First, IP identification is activated outside the device by an ID checker, while the Trojan is activated internally. For example, the Trojan can be activated by a specific event (e.g. specific input sequence, internal data value, system state) or by pre-defined timing (e.g. a specific number of clock cycles) [12, 13]. Second, the enable signal of the BFSK transmitter is provided outside the salware: it is the same signal as that used to activate the IP identification. For malware, the BFSK transmitter's enable signal is driven internally by the hardware Trojan control logic. In this case, the Trojan activates the enable signal when it is ready to send the stolen data. Note that an enable signal is required in both applications to reduce the power consumed by the ring oscillator. Moreover, a permanently activated transmitter could be detected more easily by a spectral analysis of electromagnetic emanations of the device and could also cause local heating and premature aging of the chip.

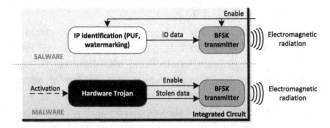

Fig. 3. Electromagnetic transmission of data (i.e. IP identification data or stolen secret data by a hardware Trojan such as the secret key for symmetric cipher).

2.2 Ultra-Lightweight Digital BFSK Transmitter

Electromagnetic radiation is an efficient side channel since, unlike measurement of power consumption, electromagnetic radiation can be measured locally. For this reason, we use the electromagnetic channel for our IP identification scheme. To this end, we designed an ultra-lightweight BFSK transmitter which could be activated outside the device by an ID checker or internally by a specific event (e.g. specific input sequence, internal data value, system state). Note that an enable signal is required to reduce the power consumed by the ring oscillator. Moreover, a permanently activated transmitter could be detected more easily by a spectral analysis of electromagnetic emanations of the device and could also cause local heating and premature aging of the chip.

BFSK is one of the common modulation schemes used in digital communication. The binary data are sent using a sinusoidal carrier at two frequency tones f_0 and f_1, representing high ('1') and low ('0') logic levels. The binary data arriving at the transmitter input at certain bitrates determine the commutation of the tones at the transmitter output. The proposed BFSK transmitter uses a dedicated configurable ring-oscillator, as shown in Fig. 4. The configurable ring-oscillator is designed using one multiplexor, $N + K$ delay elements, and a feedback chain controlled by a NAND gate for activation of transmission to reduce power consumption. Actually, the transmitter is used only during a short time when the enable signal is high, and it consumes power only during this small piece of time. The power consumption of this transmitter is thus completely negligible.

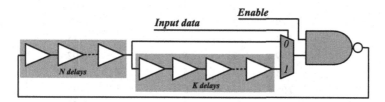

Fig. 4. Architecture of the ultra-lightweight digital BFSK transmitter based on a configurable ring oscillator.

Input data controls the multiplexor, as shown in Fig. 4. When input data is low, the ring oscillator uses N delays and its oscillation frequency is f_0. When input data is high, the ring oscillator uses $N + K$ delays and its oscillation frequency is f_1. Since the ring oscillator's oscillation frequency decreases with an increase in the number of delay elements, frequency f_0 is higher than frequency f_1. These two frequencies have to be selected according to the bandwidth of the electromagnetic analysis platform, which is used to acquire and measure the transmitted signal. The bandwidth of our test bench, which is described in Sect. 3, was limited to 100 MHz and 1 GHz by the low-noise amplifier.

The proposed BFSK transmitter was implemented in Microsemi FUSION flash based FPGA (130 nm CMOS technology) containing 600K logic gates (M7AFS600). The device contains 13 824 tiles, each tile can be used to implement one D-flip-flop or one configurable multiplexor-based logic block implementing any 3-input logic function.

The configurable number of delays in the ring oscillator of the proposed BFSK transmitter makes it possible to select precisely the two frequencies f_0 and f_1 using parameters N and K. Table 1 lists the ring oscillator frequencies and the number of Fusion tiles used by the BFSK transmitter for five values of N and K, with N ranging from 0 to 4, and K ranging from 1 to 5. According to Table 1, f_0 can be chosen between 119 MHz ($N = 4$) and 385 MHz ($N = 0$) and f_1 can be chosen between 70 MHz ($N = 4$, $K = 5$) and 280 MHz ($N = 0$, $K = 1$). The exact value of f_0 depends on the number of delay elements, but also on the placement and routing of the transmitter. For the values N and K listed in Table 1, the frequency variation was less than 1.7 % (the maximum frequency deviation in Table 1 is 2 MHz when $N = 4$).

The number of tiles used by the BFSK transmitter is very low, i.e. from 3 tiles ($N = 0$, $K = 1$) to 11 tiles ($N = 4$, $K = 5$). These values are equivalent to less than 0.022 % and less than 0.080 % of the total number of tiles included in the targeted

Table 1. Hardware implementation results of the BFSK Transmitter

N	K	f_0 (MHz)	f_1 (MHz)	Fusion Tiles	LUT4	EG
0	1	385	280	3	2	4.67
	2	383	210	4	3	5.34
	3	384	151	5	4	6.01
	4	385	130	6	5	6.68
	5	381	111	7	6	7.35
1	1	272	189	4	3	5.34
	2	272	156	5	4	6.01
	3	270	120	6	5	6.68
	4	271	106	7	6	7.35
	5	269	93	8	7	8.02
2	1	168	144	5	4	6.01
	2	169	124	6	5	6.68
	3	169	100	7	6	7.35
	4	168	91	8	7	8.02
	5	168	79	9	8	8.69
3	1	146	128	6	5	6.68
	2	147	112	7	4	7.35
	3	146	92	8	5	8.02
	4	145	84	9	6	8.69
	5	144	74	10	7	9.36
4	1	123	110	7	6	7.35
	2	121	98	8	7	8.02
	3	122	83	9	8	8.69
	4	121	77	10	9	9.36
	5	119	70	11	10	10.03

600K-gate FUSION FPGA, respectively. This very small number of tiles is very promising for good dissimulation of the BFSK transmitter inside the sea of gates/tiles. The number of FUSION tiles required by the BFSK transmitter is given by the following Eq. (1).

$$Number\_FTiles = N + K + 2 \qquad (1)$$

In order to estimate the number of resources needed for implementation with Xilinx SRAM FPGA or Altera SRAM FPGA, Table 1 gives the number of 4-input look-up-tables (LUT4) used by the BFSK transmitter with such FPGAs. The number of LUT4 required by the BFSK transmitter is given by the following Eq. (2).

$$Number\_LUT4 = N + K + 1 \qquad (2)$$

To evaluate the logical resources needed by the BFSK transmitter in ASIC implementations, the right hand column in Table 1 gives the number of equivalent gates (EG) in the transmitter. The gate count was estimated using the Virtual Silicon standard cell library based on the UMC L180 0.18 μm 1P6M Logic process (UMCL18G212T3 [31]). The delay gates are replaced by more efficient standard NOT gates. The gate count of a standard NOT gate is 0.67 EG, and that of the standard multiplexor, 2.33 EG. The standard NAND gate uses 1 EG. So the number of gates of the whole BFSK transmitter ranges from 4.67 EG ($N = 0$, $K = 1$) to 10.03 EG ($N = 4$, $K = 5$). Note that one flip-flop requires between 5.33 EG and 12.33 EG to store a single bit [31].

Such a transmitter is clearly ultra-lightweight in both FPGA and ASIC implementations. The small logical resources requirement of the proposed spy circuitry makes reverse engineering it harder, although not impossible [32]. Even with recent CMOS technologies, the attacker can reverse engineer ICs using a scanning electron microscope and an automatic tool for circuitry extraction [32, 33]. Nevertheless, the smaller the piece of hardware used for BFSK transmitter the harder it is to detect during reverse engineering. Detection of the transmitter using standard Trojan detection methods [34], [35] is not feasible because the transmitter does not change the data path of the circuit and because of the ultra-low signal-to-noise ratio on the electromagnetic channel, as shown in our experimental results below (Sect. 4).

3 Experimental Results

The electromagnetic radiation of the device was evaluated using the near-field electromagnetic analysis test bench described in [30]. The border between the far field and the near field can be considered to be about 23 mm from the device, depending on the hardware concerned. The most important part of the test bench is the acquisition chain. It determines the signal to noise ratio and measurement precision.

The chain, as presented in Fig. 5, is composed of:

- A Langer magnetic probe with a frequency range of from 30 MHz to 3 GHz and a spatial resolution of approximately 500 μm.
- A Miteq low-noise amplifier with a frequency range of from 100 MHz to 1 GHz.
- A Tektronix real time signal analyzers RSA5106B with a frequency range from 1 Hz to 6.2 GHz [36].

As presented in Fig. 5, the device to be tested (the board) is fixed to a XYZ table with repeatability of movement of 1 μm. The test bench, including the acquisition chain, XYZ table, FPGA configuration and power supply variations, is controlled by a computer. This test bench was first developed for electromagnetic attacks of TRNGs [29, 30].

The targeted FPGA for the experimental work is an Altera Cyclone III EP3C25 that uses a 65 nm CMOS technology. It contains 24 624 four-inputs LUT and 608 256 RAM bits.

Electromagnetic analysis of IC is contactless, local, and can be spatial or/and temporal. This last point makes it possible to perform frequency analysis of the electromagnetic emanation. In the your bench the spectral range is limited to 100 MHz and 1 GHz. Standard devices aimed at direct BFSK demodulation cannot be used for these

ACQUISITION CHAIN

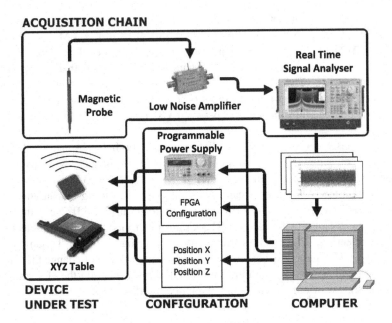

Fig. 5. Near-field electromagnetic analysis test bench

relatively high frequencies. Available integrated BFSK demodulators are limited to a few dozen megahertz. For this reason, we developed a dedicated BFSK demodulation scheme for our needs, in which a spectral analysis of the low noise amplifier output (a component of the test bench) is performed to measure the f_0 and f_1 spectral contribution. The transmitted high (low) level is detected when f_1 spectral contribution is higher (lower) than that of f_0.

For the coherent demodulation of the electromagnetic radiation, we propose a slippery window spectral analysis. Indeed, overall spectral analysis masks the effects of the no stationarity of the signal and therefore provides no information about its temporal evolution. Slippery window spectral analysis is a three-dimensional representation of the signal: amplitude, frequency, and time. It requires two quantities Fw, the width of the FFT window frame and the difference $\Delta\tau$ between two frames. For our experiment, we chose Fw equal to 16 384 points (2^{14}-point FFT) and $\Delta\tau$ equal to 100 points. For each frame, the FFT provides the software demodulator with the amplitude of signals f_0 and f_1 which enables the demodulator to distinguish between a transmitted '1' or '0'.

To illustrate data transmission from the circuit via the EM channel, we used a shift register that stored the following 16-bit sequence: "0101000111110011". The clock frequency of the shift register is 1 MHz. When the enable signal of the transmitter is given, the sequence is sent cyclically to the BFSK transmitter, which transmits it via the electromagnetic channel. The following gives the result of the coherent demodulation obtained at a 1 Mbps bit rate, which served as a proof of concept.

Figures 6 and 7 present the temporal evolution of the spectral analysis (amplitude) of the BFSK transmitter's electromagnetic emission when $N = 6$ and $K = 10$, which corresponds to the following frequencies: $f_0 = 289$ MHz (Fig. 6) and $f_1 = 119$ MHz (Fig. 7). Notice also that we placed a small antenna in the close vicinity of the ring. With $N = 6$ and $K = 10$ the BFSK transmitter uses only 17 four-inputs LUT of the FPGA that represents 0.065 % of the available logical resources of the used Altera FPGA for theses experimental results.

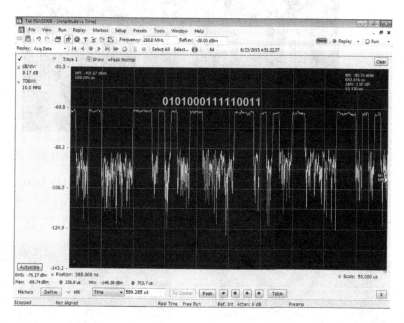

Fig. 6. Amplitude vs time evolution of the spectral analysis at $f_0 = 289$ MHz.

For the direct coherent demodulation of the electromagnetic radiation, we propose to use a slippery window spectral analysis in order to obtain a spectral cartography. Indeed, overall spectral analysis masks the effects of the non-stationarity of the signal and therefore provides no information about its temporal evolution. Slippery window spectral analysis provides a three-dimensional representation of the signal (spectral cartography): amplitude, frequency, and time. The used Tektronix real time signal analyzers [36] allows us to obtain directly the spectral cartography with direct reading of the patent that contains the transmitted data sequence. Figure 8 shows the spectral cartographies obtained at $f_0 = 289$ MHz and $f_1 = 119$ MHz.

Without knowledge of the BFSK parameters, the electromagnetic transmission cannot be easily detected because it cannot be distinguished from spectral noise. The signal-to-noise ratio of the BFSK transmission is -135 dB for a 1 GHz bandwidth. Such an ultra-low SNR represents efficient protection against unwanted BFSK transmitter detection via a side channel. However, knowing the N and K parameters, the

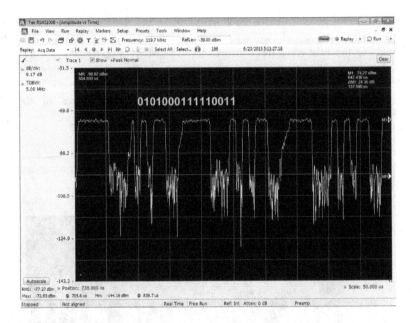

Fig. 7. Amplitude vs time evolution of the spectral analysis at f_0 = 119 MHz.

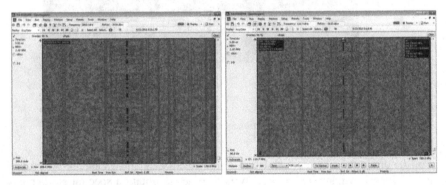

Fig. 8. Spectral cartographies center (red trace) on f_0 = 289 MHz (left) and on f_1 = 119 MHz (right) with 1 Mbps data rate. (Color figure online)

BFSK designer can calibrate the demodulation (determine the two frequencies) by electromagnetic analysis of the ring oscillators based on the differential spectral analysis as described in [30].

4   Second Version of the Ultra-Lightweight Digital EM Transmitter

Now, based on the previous experimental results we propose an enhance EM transmitter. Indeed, by using the Tektronix real time signal analyzers [36], we are able to clearly distinguish the transmitter data sequence by analyzing the spectral cartography at only one frequency. Originally, we use two frequencies in order to simplify the demodulation [11].

The new version of the transmitter is based on the use of only one frequency with an on-off controllable ring-oscillator. The binary data are sent using only one sinusoidal carrier at f_0 representing high ('1') logic level. A low ('0') logic level is obtained without transmit sinusoidal carrier (i.e. the ring-oscillator is *off*). The proposed EM transmitter uses an on-off controllable configurable ring-oscillator, as shown in Fig. 9. The on-off controllable ring-oscillator is designed using one multiplexor and K delay elements, and a feedback chain controlled by a two NAND gate for activation of transmission. The logical resource required for this transmitter is really low because it uses only K 4-inputs LUT with Altera and Xilinx FPGAs.

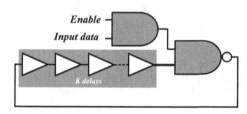

Fig. 9. Architecture of the ultra-lightweight digital EM transmitter based on an on-off controllable ring oscillator.

The previously presented (Fig. 5) experimental setup is used to test this transmitter. The parameter K is fixed to 6 to set the frequency f_0 to 309 MHz. The Fig. 10 presents the measured spectral cartographies center to f_0 with four data rates from 1 Mbps to 16 Mbps (we used the same data sequence that for the Fig. 8). This figure shows that it is possible to demodulate the transmitted sequence for 1 Mbps and 2 Mbps data rates without signal processing. It is harder to directly demodulate the data for 4 Mbps and 16 Mbps data rates, but they could be demodulated by using signal processing. These results clearly validate the proof of concept of the proposed EM transmitter. Nevertheless, the provided results (Fig. 10) are given when the EM transmitter works alone inside the FPGA.

In order to test the behavior of the EM transmitter when it is embedded in a larger system we add two large IPs in the FPGA. These two IPs are an AES cipher and a AES decipher [21]. These two IPs require 1 772 LUT4 (4.76 % of the targeted Altera FPGA) when the EM transmitter requires only 6 LUT4 (0.025 % of the targeted Altera FPGA). The transmitter ratio between the two IPs and the EM transmitter is equal to 295.3. Figure 11 presents the floorplan of the FPGA after configuration. Notice that we

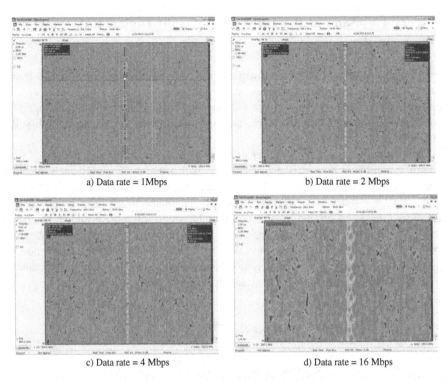

a) Data rate = 1Mbps

b) Data rate = 2 Mbps

c) Data rate = 4 Mbps

d) Data rate = 16 Mbps

Fig. 10. Spectral cartographies center (red trace) on f_0 = 309 MHz with 1 Mbps data rate (a), 2 Mbps (b), 4 Mpbs (c) and 16 Mbps (d). (Color figure online)

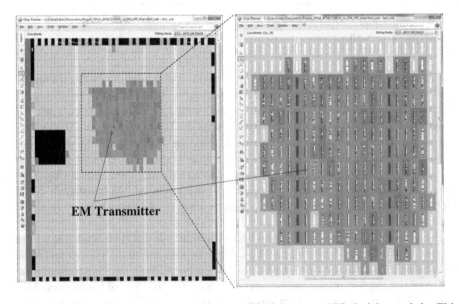

EM Transmitter

Fig. 11. Floorplan of the test system with one AES cipher, one AES decipher and the EM transmitter.

have forced the placement of the EM transmitter in order to place it in the center of the full system.

In this case, it is always possible to demodulate the data sequence with 1 Mbps data rate. Nevertheless, it could be necessary to precisely place the EM probe over the device under test. Figure 12 illustrates the modification of the spectral cartography at f_0 function of the positions of the probe. Four positions, $P_{\#0}$, $P_{\#1}$, $P_{\#2}$ and $P_{\#3}$ are tested on the same horizontal axis. The space between each position is a 3 mm gap. On Fig. 12, it appears that the data sequence is always visible, but du to other spectral contributions that come from the two other IPs inside the device, it is more difficult to directly demodulate the sequence for the positions $P_{\#0}$ and $P_{\#1}$. The position $P_{\#2}$ gives the best result. This research of the best position is really fast for the designer of the system because he knows the precise position of the EM transmitter inside the chip.

5 Comparison with State of the Art Spy Circuitries Using a Side-Channel

Table 2 compares the implementation of the proposed ultra-lightweight BFSK transmitter with other recently published state of the art methods. Table 2 gives the spy circuitry application (*App.*) for each reference; this may be IP protection (*IPP*) or hardware Trojan (*HT*) or both (for the presented work, *PW* [11]). In addition, Table 2 gives the year of publication (*YoP*), the side channel used, the hardware resources required only for the leakage generator (for example we do not take the hardware used for IP watermark generation or the Trojan's payload into account). Unfortunately, the principles compared do not use the same amount of hardware resources. For the sake of correctness, we give the implementation results as they are presented in the referenced papers. Nevertheless, the implementation bitrate of these previously published works can be roughly compared with our proposed solution. Based on published data, we computed the bitrate of all the proposals by using the number of clock cycles needed to send information via the side channel. For all the references presented in this table, the bitrate was computed using a 1 MHz frequency for data synchronization (same frequency is used during the experimental works presented previously).

As can be seen in Table 2, the proposed work reaches the highest bitrate. The reason for such a good performance is first that we use a spectral analysis of the local electromagnetic leak based on a simple frequency modulation. Except for [17], all the other solutions use a global measurement of power consumption, which reduces the signal-to-noise ratio of the information leaked via the side channel. Our proposal is clearly the smallest spy circuitry ever published. Although solutions based on circular shift-registers are well adapted to last generation FPGA families, since the 16-bit shift registers can be designed using only one look-up table, they are not suitable for ASIC technologies. Currently, an ASIC implementation of a 16-bit shift register requires 16 flip-flops whereas the solution we propose occupies an area equivalent to only one D-flip-flop.

In this chapter, we present the proposed spy circuitry for IP protection, but it can also be used for hardware Trojan. Most of the other proposals could also be used for both applications. Note that in 2012, Kasper et al. proposed to use the work initiated in [22] for hardware Trojan or IP watermarking implementation [37]. However, by using

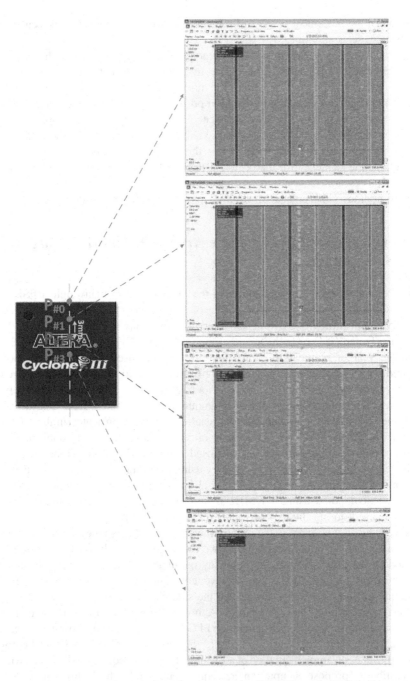

Fig. 12. Spectral cartographies center (red trace) on f_0 = 309 MHz with 1 Mbps data rate for four different positions of the EM probe over the chip. (Color figure online)

Table 2. Summary of characteristics of spy circuitries exploiting side-channels

App.	Ref.	YoP	Side channel	Hardware resources	Target	Bit rate @ 1 MHz
IPP	[17]	2008	Thermal emanation	255 Spartan-3 slices	Xilinx Spartan-3	14.10^{-3} bps
	[18]	2008	Power consumption	16 * 16-bit circular shift-registers	Xilinx Spartan-3 and Virtex-II	400 bps
	[19]	2010	Power consumption	16-bit circular shift-register	Xilinx Virtex-II Pro	1 Kbps
HT	[22]	2009	Power consumption	8 parallel D-flip-flops or 16-bit circular shit register	Xilinx Spartan-3E and Virtex-II Pro	970 bps
	[23]	2013	Power consumption	16-bit circular shift-registers per bit	Xilinx Virtex-5	1.9 kbps
Both	PW	2015	Electro-magnetic emanation	1 configurable ring-oscillator (like a D-flip-flop in ASIC)	Altera Cyclone III	1 Mbps

electromagnetic emanation and a configurable ring oscillator, the proposed solution is the most convincing for industrial applications (e.g., those aimed at IP protection) because of its very small area and high bitrate.

6 Industrial Scenarios Using the Proposed IP Protection

According to the previous section, in comparison with other works, our proposal goes clearly towards using a spy circuit in an industrial context for IP protection. Two industrial scenarios are presented in the following. The first scenario is the identification of embedded IP in the supply chain. This identification is used in order to be sure to don't use counterfeiting (fake) devices.

It is therefore crucial and strategic to be able to detect counterfeit IC as soon as possible in the supply chain (this is particularly crucial for military and space grad devices). Figure 13 shows a possible framework to manage the device under test (control the enable signal) and check the IP identification by using an EM probe, an amplifier and a dedicated acquisition system including a spectral analysis and the proposed demodulation method. Due to the high bit rate of the proposition solution the identification of the ID requires less than 500 μs (with 1 Mbps bit rate). This counterfeiting detection could be completed by other physical (invasive or not) and optical inspection [38].

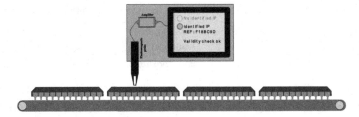

Fig. 13. Rapid and contactless IP identification in the supply chain by using EM transmission of IP' ID.

The second scenario occurs when an IP designers would like to verify the illegal presence of its IP inside a device (ASIC or FPGA). In this case the proposed transmitter provides to the identification scheme a data like a PUF [39] or a watermarking. Watermarking is a technique of steganography which provides the ownership of an IC (or an IP) by checking for the presence of hidden information called the watermark [8, 9]. Most of the watermarking methods proposed in the literature need a complex verification scheme. Nevertheless it is possible to use power consumption as proposed in [9] but it is easy and cheap to use global countermeasure in order to mask the power consumption due to the watermark [40]. Using electromagnetic emanation in this scenario is better because as electromagnetic emanation is local it is really hard to mask it by using a global countermeasure. Moreover, in this paper we have shown that due to the SNR of BFSK signal, it is unrealistic to try to detect it without the precise knowledge of the used frequencies for data transmission.

7 Conclusion

IP protection has become crucial topics for hardware security due to the lack of trust in IP market. In this chapter, we have presented two ultra-lightweight transmitters of IP identity using the electromagnetic side channel. Based on a configurable ring oscillator, our first solution exploits a BFSK signal to transmit information by way of the electromagnetic channel. Our second version is simpler; it is based on an on-off ring oscillator to transmit information with only one frequency. By performing a slippery window spectral analysis of the near field electromagnetic emanations captured locally over the transmitter circuitry, the proposed transmission achieves a high bitrate (experimentally at less 1 Mbps), which has not been achieved before. Moreover, the transmitter occupies very small area less than the requirement of a small D-flip-flop. Such a small requirement of logical resources makes reverse engineering of the chip very difficult and detection of the transmitter using standard Trojan detection methods is not feasible.

Acknowledgment. The work presented in this paper was realized in the frame of the SAL-WARE project number ANR-13-JS03-0003 supported by the French "Agence Nationale de la Recherche" and by the French "Fondation de Recherche pour l'Aéronautique et l'Espace".

References

1. Rostami, A., Koushanfar, F., Rajendran, J., Karri, R.: Hardware security: threat models and metrics. In: Proceedings of the 32nd IEEE/ACM International Conference on Computer-Aided Design, ICCAD 2013, pp. 819–823, November 2013
2. Colombier, B., Bossuet, L.: Survey of hardware protection of design data for integrated circuits and intellectual properties. Comput. Digit. Tech. IET 8(6), 274–287 (2014)
3. Ke, H., Carulli, J.M., Makris, Y.: Counterfeit electronics: a rising threat in the semiconductor manufacturing industry. In: Proceedings of the IEEE International Test Conference, ITC 2013, pp. 1–4, September 2013
4. Gorman, C.: Counterfeit chips on the rise. IEEE Spectr. 49(6), 16–17 (2012)
5. AGMA, Alliance for Gray Markets and Counterfeit Adatement. http://www.agmaglobal.org
6. Pecht, M., Tiku, S.: Bogus: electronic manufacturing and consumers confront a rising tide of counterfeit electronics. IEEE Spectr. 43(5), 37–46 (2006)
7. Bossuet, L., Hely, D.: SALWARE: salutary hardware to design trusted IC. In: Workshop on Trustworthy Manufacturing and Utilization of Secure Devices, TRUDEVICE 2013, May 2013
8. Legal, B., Bossuet, L.: Automatic low-cost IP watermarking technique based on output mark insertion. J. Des. Autom. Embed. Syst. 16(2), 71–92 (2012). Springer
9. Marchand, C., Bossuet, L., Jung, E.: IP watermark verification based on power consumption analysis. In: Proceedings of the 27th IEEE International System-on-Chip Conference, SOCC 2014, pp. 330–335 (2014)
10. Katzenbeisser, S., Kocabaş, Ü., Rožić, V., Sadeghi, A.-R., Verbauwhede, I., Wachsmann, C.: PUFs: myth, fact or busted? A security evaluation of physically unclonable functions (PUFs) cast in silicon. In: Prouff, E., Schaumont, P. (eds.) CHES 2012. LNCS, vol. 7428, pp. 283–301. Springer, Heidelberg (2012)
11. Bossuet, L., Bayon, P., Fischer, V.: Contactless transmission of intellectual property data to protect FPGAs designs. In: Proceedings of the IFIP/IEEE International Conference on Very Large Scale Integration, VLSI-SOC 2015, pp. 19–24 (2015)
12. Karri, R., Rajendran, J., Rosenfeld, K., Tehranipoor, M.: Trustworthy hardware: identifying and classifying hardware trojans. Comput. IEEE 43(10), 39–46 (2010)
13. Tehranipoor, M., Koushanfar, F.: A survey of hardware Trojan taxonomy and detection. Des. Test Comput. IEEE 27(1), 10–25 (2010)
14. Narasimhan, S., Bhunia, S., Chakraborty, R.S.: Hardware IP protection during evaluation using embedded sequential trojan. Des. Test Comput. IEEE 29(3), 70–79 (2012)
15. Kocher, P.C., Jaffe, J., Jun, B.: Differential power analysis. In: Wiener, M. (ed.) CRYPTO 1999. LNCS, vol. 1666, pp. 388–397. Springer, Heidelberg (1999)
16. Kamoun, N., Bossuet, L., Gazel, A.: Experimental implementation of 2ODPA attacks on AES design with flash-based FPGA Technology. In: Proceedings of the 22nd IEEE International Conference on Microelectronics, IMC 2010, pp. 407–410 (2010)
17. Marsh, C., Kean, T., Mclaren, D.: Protecting designs with a passive thermal tag. In: Proceedings of the 15th IEEE International Conference on Electronics, Circuits and Systems, ICECS 2008, pp. 218–221, September 2008
18. Ziener, D., Teich, J.: Power signature watermarking of IP cores for FPGAs. J. Signal Process. Syst. 51, 123–136 (2008). Springer
19. Becker, G.T., Kasper, M., Moradi, A., Paar, C.: Side-channel based watermarks for integrated circuits. In: Proceedings of the IEEE International Symposium on Hardware-Oriented Security and Trust, HOST 2010, pp. 30–35, June 2010

20. Kerckhof, S., Durvaux, F., Standaert, F.X., Gérard, B.: Intellectual property protection for FPGA designs with soft physical hash functions: first experimental results. In: Proceedings of the IEEE International Symposium on Hardware-Oriented Security and Trust, HOST 2013, pp. 7–12, June 2013

21. Bossuet, L., Grand, M., Gaspar, L., Fischer, V., Gogniat, G.: Architectures of flexible symmetric key crypto engines – a survey: from hardware coprocessor to multi-crypto-processor system on chip. ACM Comput. Surv. 45(4), 32 (2013). Article 41

22. Lin, L., Kasper, M., Güneysu, T., Paar, C., Burleson, W.: Trojan side-channels: lightweight hardware trojans through side-channel engineering. In: Clavier, C., Gaj, K. (eds.) CHES 2009. LNCS, vol. 5747, pp. 382–395. Springer, Heidelberg (2009)

23. Kutzner, S., Poschmann, A., Stöttinger, M.: TROJANUS: an ultra-lightweight side-channel leakage generator for FPGAs. In: Proceedings of International Conference on Field-Programmable Technology, ICFPT 2013, pp. 160–167, December 2013

24. Gaspar, L., Fischer, V., Bernard, F., Bossuet, L., Cotret, P.: HCrypt: a novel reconfigurable crypto-processor with secured key management. In: Proceedings of the International Conference on ReConFigurable Computing and FPGAs, ReConFig 2010, pp. 280–285 (2010)

25. Gallais, J.-F., Großschädl, J., Hanley, N., Kasper, M., Medwed, M., Regazzoni, F., Schmidt, J.-M., Tillich, S., Wójcik, M.: Hardware trojans for inducing or amplifying side-channel leakage of cryptographic software. In: Chen, L., Yung, M. (eds.) INTRUST 2010. LNCS, vol. 6802, pp. 253–270. Springer, Heidelberg (2011)

26. Du, D., Narasimhan, S., Chakraborty, R.S., Bhunia, S.: Self-referencing: a scalable side-channel approach for hardware Trojan detection. In: Mangard, S., Standaert, F.-X. (eds.) CHES 2010. LNCS, vol. 6225, pp. 173–187. Springer, Heidelberg (2010)

27. Narasimhan, S., Dongdong, D., Chakraborty, R.S., Paul, S., Wolff, F.G., Papachristou, C.A., Roy, K., Bhunia, S.: Hardware Trojan detection by multiple-parameter side-channel analysis. IEEE Trans. Comput. 62(11), 2183–2195 (2013)

28. Rad, R., Plusquellic, J., Tehranipoor, M.: A sensitivity analysis of power signal methods for detecting hardware trojans under real process and environmental conditions. IEEE Trans. Very Large Scale Integr. VLSI Syst. 18(12), 1735–1744 (2010)

29. Bayon, P., Bossuet, L., Aubert, A., Fischer, V., Poucheret, F., Robisson, B., Maurine, P.: Contactless electromagnetic active attack on ring oscillator based true random number generator. In: Schindler, W., Huss, S.A. (eds.) COSADE 2012. LNCS, vol. 7275, pp. 151–166. Springer, Heidelberg (2012)

30. Bayon, P., Bossuet, L., Aubert, A., Fischer, V.: EM leakage analysis on true random number generator: frequency and localization retrieval method. In: Proceedings of the Asia Pacific International Symposium and Exhibition on Electromagnetic Compatibility, APEMC 2013, May 2013

31. Virtual Silicon Inc. 0.18 μm VIP Standard Cell Library Tape Out Ready, Part Number: UMCL18G212T3, Process: UMC Logic 0.18 μm Generic II Technology: 0.18 μm (2004)

32. Torrance, R., James, D.: The state-of-the-art in semiconductor reverse engineering. In: Proceedings of the 48th Design Automation Conference, DAC 2011, ACM/EDAC/IEEE, pp. 333–338 (2011)

33. Subramanyan, P., Tsiskaridze, N., Li, W., Gascon, A., Tan, W., Tiwari, A., Shankar, N., Seshia, S., Malik, S.: Reverse engineering digital circuits using structural and functional analyses. IEEE Trans. Emerg. Top. Comput. 2(1), 63–80 (2013)

34. Agrawal, D., Baktir, S., Karakoyunlu, D., Rohatgi, P., Sunar, B.: Trojan detection using IC fingerprinting. In: Proceedings of the IEEE Symposium on Security and Privacy, pp. 296–310 (2007)

35. Jin, Y., Makris, Y.: Hardware Trojan detection using path delay fingerprint. In: IEEE International Workshop on Hardware-Oriented Security and Trust, HOST 2008, pp. 51–57 (2008)
36. Tektronix, RSA5000 Series, Spectrum Analyzers Datasheet (2015). http://www.tek.com/sites/tek.com/files/media/media/resources/RSA5000-Series-Spectrum-Analyzers-Datasheet-37W2627414_1.pdf
37. Kasper, M., Moradi, A., Becker, G.T., Mischke, O., Güneysu, T., Paar, C., Burleson, W.: Side channels as building blocks. J. Cryptogr. Eng. 2(3), 143–159 (2012). Springer
38. Tehranipoor, M., Guin, U., Forte, D.: Counterfeit Integrated Circuits - Detection and Avoidance. Springer, Heidelberg (2015)
39. Bossuet, L., Ngo, X.T., Cherif, Z., Fischer, V.: A PUF based on a transient effect ring oscillator and insensitive to locking phenomenon. IEEE Trans. Emerg. Top. Comput. 2(1), 30–36 (2014)
40. Kamoun, N., Bossuet, L., Ghazel, A.: Correlated power noise generator as a low cost DPA countermeasure to secure hardware AES cipher. In: Proceedings of the International Conference on Signals, Circuits and Systems, SCS 2009, pp. 1–6 (2009)

JAIP-MP: A Four-Core Java Application Processor for Embedded Systems

Chun-Jen Tsai[✉], Tsung-Han Wu, Hung-Cheng Su, and Cheng-Yang Chen

Department of Computer Science, National Chiao Tung University, Hsinchu, Taiwan
cjtsai@cs.nctu.edu.tw

Abstract. In this chapter, we present a four-core Java application processor, JAIP-MP. In addition to supporting multi-core coherent data accesses to shared memory, each processor core in JAIP-MP is a hardwired Java core that is capable of dynamic class loading, two-fold bytecode execution, object-oriented dynamic resolution, method/object caching, Java exception handling, preemptive multi-threading, and memory management. Most of the essential OS kernel functions are implemented in hardware. In particular, the preemptive multi-threading performance is much higher than that of a software-based VM running on a traditional OS kernel such as Linux. Currently, single-cycle context switching with a time quantum as small as 20 µs can be achieved by each core. More importantly, the Java language model itself makes it possible to maintain binary portability of application software regardless of the hardwired OS kernel component. In summary, JAIP-MP could be used to study the potential benefits of OS kernel implementation in hardware.

Keywords: Java processors · Multi-core processors · Embedded SoC · Hardwired operating system kernel

1 Introduction

The Java programming language has been one of the most popular programming languages for over a decade. There are many reasons for its popularity. For example, it is a clean language designed with object-orientated paradigm from scratch, without unnecessary features such as multiple inheritance or pointer arithmetic that can be easily abused by programmers. Memory management in Java is implied by the object-oriented model and requires no special treatment from the programmers. It maintains a great Job on backward compatibility that application binaries written for very old version of Java can usually be executed under the latest versions of Java Runtime Environment (JRE) regardless of the underneath operating systems. One of the reasons that a Java program can be portable across versions and platforms is that the Java language model defines some interfaces, such as multi-thread programming and memory management, which are usually defined by the operating systems.

There are many variations of JRE for embedded systems, including Sun's CDC/PBP, CLDC/MIDP and Google's Android platform. Most existing implementations are

© IFIP International Federation for Information Processing 2016
Published by Springer International Publishing AG 2016. All Rights Reserved
Y. Shin et al. (Eds.): VLSI-SoC 2015, IFIP AICT 483, pp. 170–192, 2016.
DOI: 10.1007/978-3-319-46097-0_9

software-centric, which means they require a sophisticated operating system (OS) to support the JRE (e.g., the reference implementations of the Java ME and the Android platforms heavily depends on Linux or OS's with similar capabilities). In general, operating systems handle thread management, memory management, I/O interfaces, and dynamic class loading from local and/or remote file systems for the JRE. However, most Java middleware stacks of JREs have already included main functions of a typical OS kernel. Therefore, adopting a complete OS underneath a JRE is a duplication of system functions, which is not a good design philosophy for embedded devices with resource constraints. A different design approach, the JavaOS model [1, 2], was proposed to implement the Java application platform. In this approach, the OS itself is written in the Java language as part of the JRE. The JavaOS system model uses a small microkernel written in native codes to support low-level physical resource management.

In this chapter, we will present the design of a multi-core Java System-On-a-Chip (SoC) that implements most of the OS kernel functions, as well as the bytecode execution pipeline using hardware circuits. The key component of the Java SoC is the Java Application IP (JAIP). JAIP is a reusable processor IP that has hardwired support of the following Java language features [3, 4]:

- Two-fold instruction folding
- Multithreading, synchronization, and exception handling
- Heap management and garbage collection
- Class inheritance and interface invocations
- Method and heap object caching
- Dynamic class loading and dynamic symbol resolution
- String operation acceleration for matching and copying

The remainder of this chapter is organized as follows. For the rest of Sect. 1, we will first introduce some previous work on multi-core Java processor designs and discuss the possible benefits of implementing a multi-core Java SoC with the major OS kernel functions crafted in hardware. Section 2 presents the architecture of the JAIP core, including the microarchitecture of the instruction execution pipeline, the stack architecture, the preemptive thread manger, and the memory manager and garbage collector. We also have a brief discussion on how the I/O subsystem interface can be merged into the dynamic resolution module of a Java VM. Section 3 discusses the required glue logics to integrate four JAIP cores into a single SoC, mainly, the inter-core communication unit, the multi-core thread manager, and the data coherence controller. The experimental results are presented in Sect. 4. Finally, some discussions are given in Sect. 5.

1.1 Multi-core Java Processors

Although there are many hardware Java core designs [5], very few of them have been synthesized in a real multi-core application processor [6]. One of the reasons may be due to the fact that previous work shows that a Just-in-time (JIT) based VM running on a high performance general-purpose processor can often outperform a hardwired Java processor [7]. Therefore, it seems that there is no need to further pursue the development of hardwired Java processors. However, most of the JIT vs. hardwired VM comparisons are conducted using benchmarks where the application class files are not optimized for Java processors. For example, it has been shown that some popular benchmark class files can run much faster on a Java processor if bytecode optimizations in the class files are conducted [8]. Please note that an optimized Java class file still conforms to the Java specification and is portable across different Java platforms. In addition, many benchmarking processes discard the impact of the JIT compilation overhead [3]. Although ignoring the JIT overhead is reasonable for some applications, it is not valid for remote invocations that are common for object-oriented distributed computing. Other reasons why a hardwired VM could be useful for practical applications will be discussed in Sect. 1.2.

Most Java processors support thread synchronization using software modules [9–11]. However, the execution time of a software-based synchronization operation, such as a mutex lock, can take more than a few hundred clock cycles since the lock objects are often accessed in conventional trap routines. PicoJava [10] uses a few special-purpose registers for the speedup of synchronization operations, but it still needs to use the main memory to maintain the information of all waiting threads and lock objects. Therefore, a high number of concurrent synchronized read/write operations can have significant synchronization overheads. JOP-CMP [6] supports at most 8 processor cores with a software-based thread scheduler and a hardwired synchronization unit [6, 12]. There is only one global lock register in the synchronization unit, which means that any threads trying to acquire the lock must wait until the lock is released. In addition, JOP-CMP does not have a coherent data cache. All data accesses will be directly issued to the external memory, which can hinder the multi-thread performance significantly.

1.2 Potentials of Hardwired Virtual Machines

Traditionally, Java programs are executed using software-based virtual machines. To improve the performance of bytecode execution, JIT or ahead-of-time (AOT) compilation techniques are often adopted in modern virtual machines. Previous work shows that JIT or AOT techniques can arguably achieve better performance than a hardwired Java processor. We have already presented some reasons that may lead to the bias of such conclusions in Sect. 1.1. Another reason is that existing Java processors mainly focuses on the architecture design of the bytecode execution pipeline. Things may be different when the full JRE is considered as the target of hardware design.

For example, one of the most intriguing features of the Java programming model is that all data accesses and code invocations must go through the dynamic resolution mechanism. Although dynamic resolution is usually considered as a language feature that hinders efficiency significantly, there are some benefits of dynamic resolution that has not been investigated thoroughly, especially when the Java VM is implemented in hardware. First of all, with dynamic resolution, there is no need to assume a large "flat" memory model for a Java VM implementation. A Java VM may manage many concurrently accessible memory blocks seamlessly to improve the performance of data processing without the programmer knowing the physical layout of the memory subsystem. Securities issues related to malicious pointer-based indirect data accesses can be handled more rigorously with the Java model since all data accesses must be approved by the dynamic symbol resolution unit (DSRU). Finally, a method call can be re-directed to hardware logics without going through memory-mapping process and shared bus transactions, which may improve throughput significantly. The last point will be explained further in Sect. 2.5.

In [4], we have presented the preemptive multi-threading efficiency of the JAIP core. It is capable of single-cycle multi-thread context switching with a time quantum as small as 20 μs. For a traditional OS kernel such as Linux, the time quantum for a thread is usually around 10 ms. As a result, for single-core multithread applications, hardware-based thread manager can achieve much smoother concurrent executions of all threads of equal priorities than a software-based thread manager. This is a very strong reason for the development of an efficient hardwired Java processor core. The Java language specification defines standard programming interfaces for OS kernel services such as process management and memory management. Other popular languages such as C and C++ do not standardize these functions. For example, thread creation API's are OS-dependent for C programs. Therefore, it would be beneficial to investigate the potentials of a fully hardwired OS kernel based on the Java programming model.

In short, a Java processor can be designed to implement the entire OS kernel system services in hardware while maintaining application portability. This is particularly important for embedded real-time applications where context-switching efficiency and dynamic memory management overhead are the key performance factors of a system. On the other hand, it is not so easy to "harden" the OS kernel for a traditional RISC processor due to lack of "standard" system calls for C/C++ applications.

2   The Architecture of the JAIP Core

In this section, we present some design details of the JAIP core that is used as the key component of the hardwired multi-core Java runtime environment (JRE). The design target of the JAIP core is for FPGAs and thus dual-port SRAM blocks that are common in FPGAs are used extensively to optimize the architecture for the object-oriented language model that makes Java one of the most popular programming languages.

2.1   The Overview of JAIP Core

Figure 1 shows the overall block diagram of a single-core SoC based on the JAIP core. The complete SoC is composed of a RISC core and a JAIP core. For the execution of a Java program, the RISC core is only responsible for reading and parsing of the class files stored in a JAR file on the Compact Flash (CF) card. The RISC parser will convert the standard Java class files into runtime class images on-the-fly for direct execution by the JAIP core. The converted class file images are stored in the second-level method area in the main memory. The class file parser will maintain a symbol cross reference table stored in the main memory for all loaded classes. The Java core is completely responsible for the two-fold execution of bytecodes, dynamic loading of the class images into the method area, dynamic resolution, memory management, and preemptive multi-thread scheduling. In the future, we will remove the dependency of the JAIP core on the RISC core for class file reading and parsing.

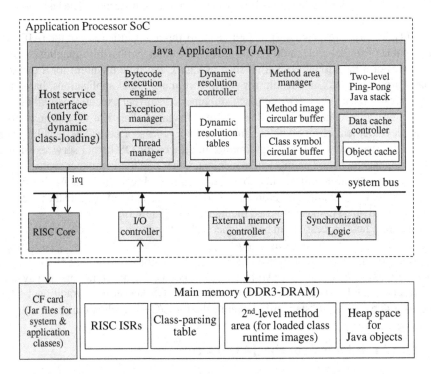

Fig. 1. The architecture of a Java application processor based on the JAIP core.

JAIP adopts a two-level method area design. All classes loaded at runtime will be stored in the main memory (i.e., the second-level method area) using the late-resolution policy. A Java method (and its related symbol information) must be loaded into the on-chip method cache (the first-level method area) before it can be executed by the bytecode execution engine. In short, the complete class images of the Java applications are stored in the main memory while the most recently used methods and symbol information are stored in the on-chip method area in a FIFO manner.

Since the Java VM is basically a stack machine, i.e., all the local variables and the intermediate values of operations are stored in the runtime stack, fast accesses to the most recent stack frames are essential to the performance of a Java processor. JAIP uses a special-purpose on-chip memory and three top-of-stack registers to form a two-level Java runtime stack. The special-purpose on-chip memory is a customized four-port memory device custom-designed for the Java bytecode instruction set architecture. It is composed of a pair of interleaving two-port memory blocks and four registers. The design is a good tradeoff between performance and implementation cost as compared to the Java processors with a large stack cache [10, 13, 14].

The two-level Java stack allows JAIP to perform two-fold instruction folding for frequent bytecode pairs [15] such as load-load, ALU-store, etc. However, to simplify the microarchitecture, some folding patterns, e.g. ALU-ALU bytecode pairs, are not allowed. According to our empirical studies, the instruction folding rate of JAIP ranges from 10 % to 40 % for different benchmark applications.

2.2   The Bytecode Execution Engine and the Stack Memory

The bytecode execution pipeline of the JAIP core is shown in Fig. 2. The Java bytecodes are translated into native JAIP instructions called j-codes before instruction decoding and folding. The JAIP core performs two-fold instruction folding of stack-related Java operations using a simple decision policy. In short, JAIP only supports the folding of the following stack operation pairs: Load-Store, Store-Load, ALU-Load, Load-ALU, Store-ALU, ALU-Store, Load-Load, and Store-Store. Note that in these stack operations, 'Load' means loading a data item on to the operand stack. The source of the data can be from the local variable area of the Java stack or a constant value. 'Store' means removing a data item from the operand stack. The destination of the removed data can be the local variable area or a null space (as in the 'pop' operation). Finally, 'ALU' means an arithmetic and logic operation. The fetch stage of the pipeline will guarantee that, at any given cycle, the j-code information passed to the decode stage belongs to one of the following three cases: a foldable j-code instruction pair, a single control instruction (such as a conditional branch), or a special data-processing j-code (such as the 'swap' operation). The two-level JAIP stack can encounter structure hazard whenever the j-code instruction pairs try to transfer two local variables stored in the same SRAM bank to the operand stack (or vice versa). This hazard can be removed by using a general-purpose four-port memory for the second-level stack. However, since a general-purpose four-port memory is often expensive, we use a special-purpose 4-port memory customized to the Java ISA to reduce the occurrence of structure hazards while maintaining low implementation cost [16].

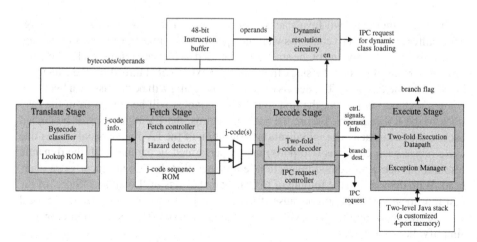

Fig. 2. The bytecode execution engine pipeline of JAIP.

According to the Java VM specification [17], the first four local variables should be the most frequently used ones (which can be arranged by an optimized Java compiler). Hence, some Java instructions (with no operands) are designed specifically for accessing these variables. The second-level Java stack memory is constructed by using two on-chip memory blocks organized in an interleaving structure to form a Java stack. In addition, four 32-bit local variable (LV) registers are used as a small cache for the first four local variables as shown in Fig. 3.

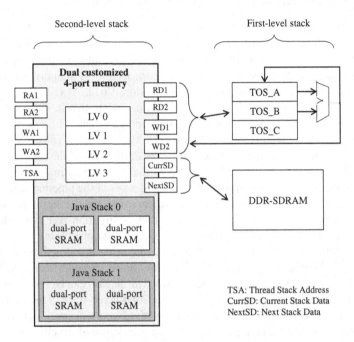

Fig. 3.   The architecture of the stacks in a JAIP core.

In Fig. 3, there are two Java stacks instead of one. These two stacks form a ping-pong buffer to support fast context-switching operations for preemptive multi-threading. At any given time, only one of the stacks will be used as the active Java stack. The other stack will be used to load the stack frame of the next selected thread for the execution in the next time quantum. The details of context switching will be discussed in Sect. 2.3. Upon a method invocation, the first four local variables will be copied from the Java stack to the LV registers. Before the method returns, the LV registers will be copied back to the Java stack. The initialization/restoration of the LV registers only takes one cycle (since each bank has two ports) and is performed in parallel with the dynamic resolution process of method invocation/return such that they do not incur extra overhead. With this design, the folding of two stack operations of 'Load' and 'Store' of the first four local variables do not cause structure hazard. However, accesses to the local variables beyond the first four will not be folded by JAIP. This is a design choice to simplify the control logic.

2.3   Single-Core Preemptive Thread Management

For the execution of multi-thread Java programs, each thread must maintain its own registers and runtime stack. Typically, the register file of a Java processor is only composed of few special-purpose registers and can be swapped out to main memory quite efficiently. On the other hand, the Java runtime stack is much larger than the register file. If the runtime stack is stored in the main memory (e.g., DRAM), there is no need to save the runtime stack. However, most high-performance Java processors, including JAIP, use stack cache or on-chip memory to support instruction folding and to reduce the access delay of operands. In either case, the time it takes to swap out the on-chip stack would be non-negligible.

Saving/restoring the context of a JAIP thread involves transferring the stack frames (each ranging from a few bytes to a few hundreds bytes) to/from the main memory. In order for JAIP to support hardware-based multi-threading, we have designed a low-cost thread manager unit to reduce the context-switching overhead. As a result, in most cases, switching from the current thread to the next active thread only takes a single cycle. This is much faster than any software-based preemptive multi-tasking operations where a context-switching operation can take anywhere from a few hundreds to over a thousand cycles.

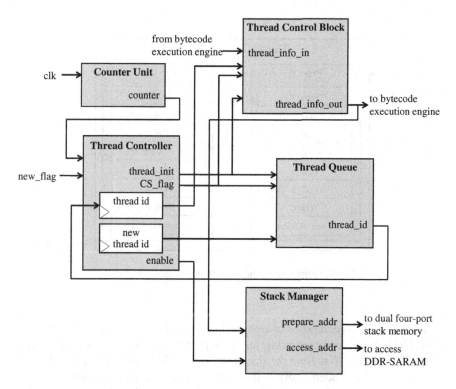

Fig. 4. The thread manager unit.

The architecture of the thread manager unit is shown in Fig. 4. When a Java program executes the start() method of an object derived from the Thread class, the JAIP execution pipeline will send a signal to the thread manager unit, informing the controller to initialize a new task in the on-chip thread control block (TCB) and enters the thread ID into the thread queue. Note that the execution of the start() method via the 'inovkevirtual' bytecode goes through the dynamic resolution unit of JAIP, which trigger the controller circuit of the thread management unit directly. More discussion on the direct invocation of hardware logic or I/O devices through a standard Java method invocation mechanism will be discussed in Sect. 2.5.

The structure of the TCB is shown in Fig. 5. In the current design, a fair round-robin algorithm is used in the controller to select the next ready thread. The state of a thread is stored in a TCB entry, which is composed of the following information:

1. The ID of the thread.
2. The Java class and method IDs of the thread.
3. The local variable pointer and the operand stack pointer.
4. The program counter and the number of local variables of the thread.
5. The first-level operand stack (the top-of-stack A, B, and C registers) of the thread.
6. The object reference (pointer) to the thread object in the Java heap.

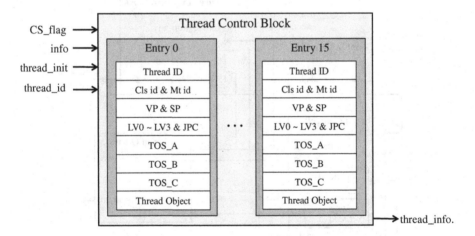

Fig. 5. Structure of the on-chip thread control block.

Each TCB entry is composed of eight 32-bit values. In the current design, the thread control block is implemented using an on-chip memory. We have set the maximal number of threads to 16 to limit the size of the TCB to 512 byte. The maximal number of threads can be extended easily at the cost of a larger on-chip memory. For thread management, we use a circular queue to store the ID of each thread in the queue. Every time a new thread is created by the Java application through the execution of the start()

method of a Thread object, a new thread ID will be generated and entered into the end of the thread queue. When the time slice of the current thread ends, its ID will be moved to the end of the queue and the thread whose ID is pointed to by the 'next' pointer will become the current thread.

The ping-pong stack architecture works as follows. As soon as a thread is selected as the current thread and starts its execution, the multi-threading logic also picks the next thread to be executed and, while the first thread is running, swaps in the runtime stack of the second thread from the main memory. When the time slice of the first thread is up, JAIP can be switched to the second thread within a cycle since its stack has already been setup. In the rare case where the restoration of the runtime stack of the second thread takes longer than the predetermined time quantum of the first thread, the time quantum will be extended until the runtime stack of the second thread is in place. The average time it takes to backup or restore a runtime stack to/from the backing store (the main memory) for the target system used in this chapter (Xilinx ML 605) is less than 10 μs when the system clock is 83.3 MHz.

When the execution is switched to the second thread, the runtime stack of the first thread will be saved to the main memory in parallel to the execution of the second thread. As soon as the stack of the first thread is saved, the multi-thread control logic will proceed to the setup of the third thread. With this design, the overhead of saving/restoring the runtime stack can be overlapped with the execution of the current thread. According to our experiments, the time slice of the proposed architecture can be as small as 20 μs and the only overhead in context-switching is virtually the reset of the processor pipeline (similar to a branch instruction). Smaller time slice means the distribution of the CPU resources to each thread is more even. This level of multi-threading efficiency is very difficult to achieve with a software-based preemptive multitasking operating system.

2.4   The Memory Manager and Garbage Collector

Garbage collection (GC) is an important feature of the Java programming model. It takes the burden of memory management off the programmers and removes common memory-related bugs in programs. However, runtime GC may induce high overhead and affect the performance of an application [18, 19]. This is particular true for software-based VM. Therefore, for embedded applications, programmers must be careful to avoid triggering GC unintentionally or the whole application may stall until the GC process is finished.

On the other hand, for hardwired Java VM, the GC circuitry can run in parallel to the bytecode execution pipeline, it is possible to design hardware based GC that does not stall the execution of the Java applications [20]. Although hardware-based GC is an active research direction [20–22], most designs are simply technical investigations and have not been integrated into a complete Java system. For example, [20] presents a synthesizable GC hardware, but the GC is exclusively evaluated on an FPGA using test patterns that represent typical applications.

Although GC is a crucial component of a JVM, the JVM specification does not enforce of any type of GCs for memory management. The memory manager hardware in JAIP includes hardware controllers that handle memory allocation and object caching (see Fig. 6). To perform garbage collection, the VM must carry out two types of operations. First, the VM must be able to determine that an object on the heap is not pointed to by any Java reference variables. Secondly, the GC mechanism must return the object space to the unused memory block list and merging two consecutive unused memory blocks if possible. In JAIP, we adopt the tracing garbage collector since it has low overhead and is suitable for hardware implementation. Furthermore, it can be a pluggable component to existing memory manager of JAIP.

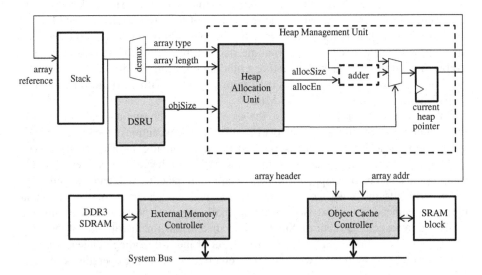

Fig. 6. The memory manager architecture of JAIP without the GC.

In short, the tracing collector returns all the local references to the unused memory block list unless the reference is a return value to the caller method. To achieve this goal, we expand the heap allocation unit in Fig. 6 to the architecture in Fig. 7. The object allocation controller is responsible for allocation of a new object on the heap and enters the object into the on-chip GC table. The GC table can be accessed by the GC controller for unused object collection upon the return of a method. Note that to hide the overhead of the GC, the GC controller must be able to access the GC table concurrently to the operation of the object allocation controller. Hence, we use a two-port memory for the GC table. Another on-chip memory in Fig. 7 is the GC method stack memory. The GC controller maintains this memory exclusively. During the execution of a method, this memory records all objects allocated locally and whether they are assigned to references outside the scope of this method. Upon the return of the method, the GC controller will go through the list of objects and return the memory blocks to the unused memory list if possible. Note that the collection process is executed in parallel to the normal bytecode execution pipeline.

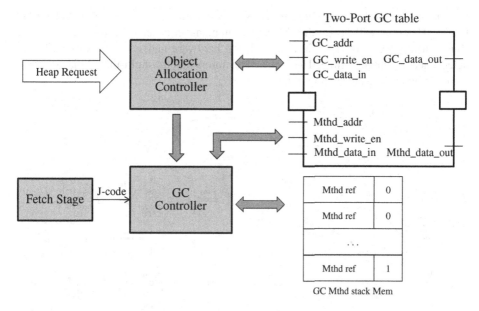

Fig. 7. The garbage collector of JAIP.

The GC controller will merges consecutive free memory block in the GC table into a larger block. However, it does not move the occupied memory blocks to create larger unused blocks because the cost would be too high for embedded applications. Note that the GC algorithm used in JAIP is not a complete garbage collector. It only collects unreferenced objects created by a method upon the return of the method to the caller. The reason this algorithm is chosen is mainly because it has very low runtime overhead and can be integrated into the existing memory manager of JAIP without major modification to the overall microarchitecture.

2.5 Dynamic Symbol Resolution Unit and the I/O Subsystem

In Sect. 1.2, we mentioned that the DSRU can provide a direct interface to the I/O subsystem of a hardwired Java VM. In this subsection, we use the JAIP DSRU as an example to explain the details. Since most modern operating systems and processors adopt the memory-mapped I/O model to manage I/O devices and accelerators, naturally, accesses to I/O devices are achieved using memory read/write operations in the I/O subsystem address space. Java uses the symbol space realized by the DSRU to replace the concept of the address space. Therefore, for a hardwired Java VM, the I/O subsystem can be integrated seamlessly into the DSRU logic. A method call in Java can be transform directly by the DSRU into control signals wired to a hardware device through some routing box (similar to the interconnect module of the ARM AXI bus protocols). Figure 8 shows the state-diagram of the controller of the DSRU of JAIP. When a program invokes a method, the controller begins at the 'IDLE' state and begins the symbol resolution process. When the DSRU determines that the target of the method invocation is for a native method implemented in hardware, it will enter the state of 'Invoke HW Logic.' This state will trigger the I/O subsystem manager to send appropriate hardware signals to the target device. Currently, all the hardware native methods of JAIP are determined at synthesis time. The string accelerators and the multi-thread managers of JAIP are invoked using such facility.

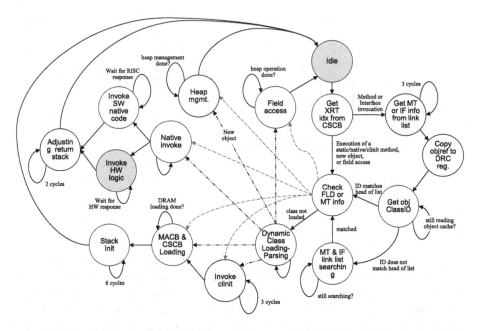

Fig. 8. The state diagram of the dynamic resolution controller of JAIP. XRT stands for 'cross-reference table,' MT stands for 'method,' IF stands for 'interface,' DRC stands for 'dynamic resolution controller,' and FLD stands for 'field.'

3   Multi-core Integration of JAIP

3.1   The Multi-core Thread Manager

In order to integrate multiple JAIP cores into one application processor, we must modify the microarchitecture of JAIP. The multi-core capable JAIP core is shown in Fig. 9. The new addition to the original JAIP core is the Inter-Core Communication Unit (ICCU). The interactions between various components of the JAIP core and the ICCU are illustrated in Fig. 10. In the Java programming language, an object belongs to the "Thread" class can register its own execution context by invocation of the Thread.start() method. At runtime, the Dynamic Symbol Resolution Unit (DSRU) of JAIP will resolve the method invocation of start() and trigger a hardwired signal to the thread manager unit of the local JAIP core that executes the start() method. Such direct invocation of a hard-wired logic through the dynamic resolution unit is called the Hardware Native Interface (HNI). In the original single-core JAIP, the local thread manager will handle the thread creation requests by itself and register a new entry in its local task queue. However, for a multi-core capable JAIP, the thread creation request cannot be handled locally. Instead, the request will trigger the HNI invocation of the ICCU, and the request signal will be passed to the Data Coherence Controller (DCC). The DCC then talks to a global thread manager to request for the creation of a new thread. The global thread manager will assign the new thread to a JAIP core based on the depth of its local task queue.

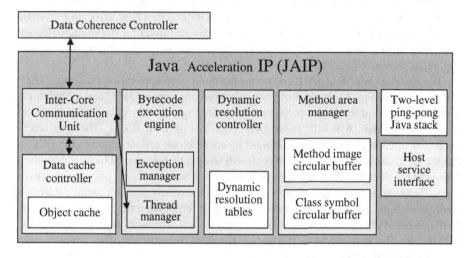

Fig. 9. Modifications required to a JAIP core to enable multi-core integration.

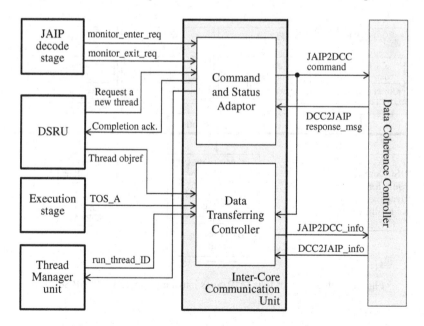

Fig. 10. Signaling between ICCU and other components of JAIP.

In addition to thread creations, the Java language also defines standard ways for synchronization. In short, each Java object contains a lock (similar to mutex in other programming language). Synchronization can be achieved explicitly through the acquisition of the lock in an object, or implicitly through invocation of a synchronized method. Similar to the thread creation problem, the acquisition of a lock cannot be handled locally since two threads requesting the same lock may be running on different JAIP cores. Therefore, such locking requests will also be passed to the ICCU for multi-core mutex operations. However, this time, the ICCU is not activated by a HNI invocation from the DSRU because the lock request is triggered by the execution of a "monitor" bytecode. Therefore, the lock request is originated from the decode stage of the bytecode execution engine, as shown in Fig. 10.

The integration of four JAIP cores into the multi-core application processor, JAIP-MP, is shown in Fig. 11. In the SoC, we only need one copy of DCC and global thread manager. The combination of these two hardware logic is referred to as the multicore coordinator of the JAIP-MP. Each JAIP core has its own ICCU. The local cache controller of each JAIP core will forward its cache block update status to the DCC so that the DCC can inform other cache controller to update their cache blocks if necessary. This is an efficient way to guarantee cache coherence when there are only few processor cores. However, to simplify the implementation of the coherent object cache, each cache controller adopts a write-through policy. This is different from the original single-core JAIP presented in [3], where a write-back policy is used. The write-through cache policy does hinder the single-core performance slightly. Nevertheless, the overall system performance still scales up fairly well.

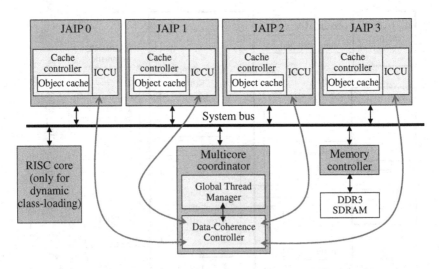

Fig. 11. Integration of four JAIP cores into a multi-core JAIP-MP SoC.

3.2   The Data Coherence Controller Architecture

The detail architecture of the DCC is shown in Fig. 12. It is composed of four sub-modules. The cache coherence controller maintains the data consistency across the object heap controllers of each core. The heap controller adopts the least-recently used policy and write-through strategy for caching of Java heap objects. The mutex controller serially decodes requests sent by the JAIP cores and activates corresponding sub-module. The thread assignment controller (TAC) is responsible for load balancing among all JAIP cores. When a JAIP core invokes the Thread.start() method, the TAC will forward its special-purpose registers to the JAIP cores with the least number of ready threads. The Lock Object Accessing Controller (LOAC) shown in Fig. 13 maintains the information of waiting threads associated with each occupied lock object.

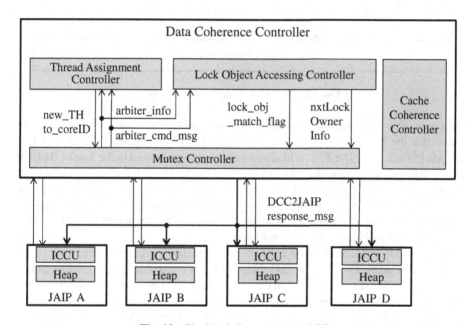

Fig. 12.   The block diagram of the DCC.

When several JAIP cores try to request locks on the same mutex concurrently, the mutex controller uses a fixed-priority policy to determine which core can lock the mutex. Currently, the JAIP core with a smaller ID has a higher priority. The mutex controller supports three types of requests: dispatching a new thread, acquiring a lock object, and releasing a lock object. Either the TAC or the LOAC will be activated after the mutex controller determines the type of the request.

When any of the JAIP cores issues a request for the dispatching of a new thread, the TAC should determine a JAIP core to handle the new thread. In order to determine the

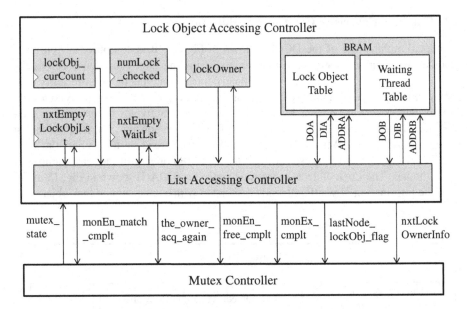

Fig. 13. The block diagram of the LOAC.

current number of active threads in each JAIP core, the TAC maintains a table. The table indexed is the ID of the JAIP core, and its entries store the current number of active threads of each core. The TAC will always assign the new thread to the lowest ID JAIP core that has the fewest number of ready threads. The TAC will inform the MHC to send a response signal to the chosen JAIP core with some essential information of the new thread. The ICCU of the JAIP core may process the information by decoding the response signal. Finally, the ICCU activates the thread manager unit of the JAIP core to add the new thread into its local thread queue.

Figure 14 is an example of the link lists maintained by the LOAC, which consists of a lock object table, a waiting thread table, and a few internal registers. Each occupied lock object maintains a linked list in these two tables. The head node of the linked list of a lock object begins at an entry in the lock object table, and the rest of the linked list nodes are entries of the waiting thread list. Each entry in the link list (except for the head node) represents a thread that is performing a lock operation on the object. The first thread in the linked list is the link list is the current owner of the lock. As soon as any thread in one of the JAIP cores tries to lock a Java object, the mutex controller will send a lock object L_n to the LOAC. The LOAC will look for the object address of an entry that matches L_n in the lock object table. Once the matched entry is found, the information must be recorded in the waiting thread table. Each entry contains the IDs of the JAIP core and the thread. The new entry is appended at the end of the link list. If the request from a thread is to release the lock object L_n, the LOAC will remove the thread from the link list. If any other thread is waiting for the same lock object L_n, the LOAC will make the second thread in the link list become the current owner of the lock object.

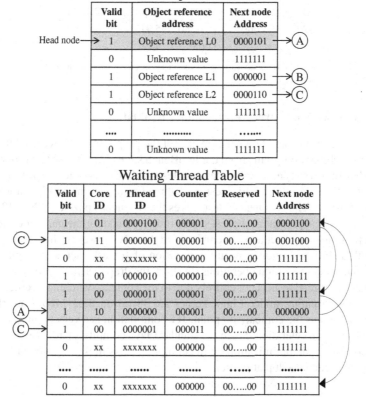

Lock Object Table

	Valid bit	Object reference address	Next node Address	
Head node →	1	Object reference L0	0000101	→ Ⓐ
	0	Unknown value	1111111	
	1	Object reference L1	0000001	→ Ⓑ
	1	Object reference L2	0000110	→ Ⓒ
	0	Unknown value	1111111	
	••••	••••••••••	•••••••	
	0	Unknown value	1111111	

Waiting Thread Table

	Valid bit	Core ID	Thread ID	Counter	Reserved	Next node Address
	1	01	0000100	000001	00.....00	0000100
Ⓒ →	1	11	0000001	000001	00.....00	0001000
	0	xx	xxxxxxx	000000	00.....00	1111111
	1	00	0000010	000001	00.....00	1111111
	1	00	0000011	000001	00.....00	1111111
Ⓐ →	1	10	0000000	000001	00.....00	0000000
Ⓒ →	1	00	0000001	000011	00.....00	1111111
	0	xx	xxxxxxx	000000	00.....00	1111111
	••••	••••••	••••••	•••••••	••••••	•••••••
	0	xx	xxxxxxx	000000	00.....00	1111111

Fig. 14. Data structures maintained by the LOAC.

4   Experimental Results

The proposed architecture has been implemented on a Xilinx ML605 platform with a Xilinx Virtex6 XC6VLX240T FPGA. The RTL model of the JAIP core and the DCC logic are written in VHDL. Four JAIP cores and one DCC logic are integrated into the application processor using Xilinx XPS 13.4. The synthesis tool is Xilinx XST 13.4 and the target clock is 83.3 MHz. According to the place-and-route timing report of the Xilinx tools, the critical path of the system is currently at the execution stage of JAIP, from the customized four-port stack memory to ALU and then back to the four-port memory. The target frequency is chosen at 83.3 MHz due to some restrictions for DDR DRAM support on the development boards. The FPGA resource usages of JAIP and DCC are shown in Table 1.

Table 1. Logic usage of a JAIP core and DCC on a Virtex6 FPGA device.

FPGA logic units	LUT6s	Flip-flops	BRAMs
JAIP (per core)	12,580	5,912	34
DCC	663	449	1

Note: LUT6 means a six-input lookup table in a logic cell of a Xilinx device.

4.1 Single-Core Multithread Performance Evaluation

To evaluate the multi-threading performance of the proposed JAIP, we used the multi-threading benchmark programs from the JemBench suites [23]. These test programs are explained as follows. The 'Dummy' test creates multiple threads to execute busy loops for 5000 iterations. For the 'Matrix' test, each thread computes the multiplication of two 20-by-20 matrices. The 'N-Queens' test solves the N-Queens puzzle for N = 13 in each thread. For each test programs, the test scores roughly stand for the number of iterations each test program can execute per seconds by all threads. However, the scores are associated with quantization noises from the partition of subtasks across multiple threads and from the synchronization operations. In short, the drop in scores from single-thread test to multiple-thread tests is not entirely due to the context-switching overhead. Sun's CVM-JIT [24] running under Linux kernel 2.6.38 on an 83.3 MHz PowerPC 405 processor is used as the comparison point. JIT compilation is a very popular technique for Java program acceleration. Since the standard Java compilers (from Sun/Oracle) do not perform bytecode optimization on the compiled class files, a JIT-based VM could achieve significant speedup at runtime.

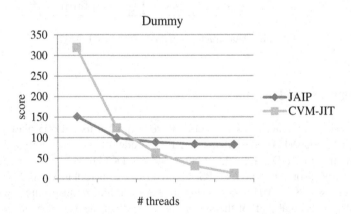

Fig. 15. JemBench scores of the 'Dummy' test on a single JAIP core.

From Figs. 15 and 16, one can see that the performance of CVM-JIT is higher for single-thread execution of the Dummy test and the Matrix test, JAIP has better performance when the number of threads becomes larger. Since in these tests, both JAIP and CVM-JIT executes using only one processor core, the scores drops naturally as the number of threads increases due to task division and synchronization overheads explained before. However, from these plots, it is quite clear that a software-based multithread mechanism such as the CVM-JIT has higher overhead in thread management. The performance drops significantly as the number of threads increases. For the Dummy test, JAIP outperforms CVM-JIT when the thread number is larger or equal to 4. For the Matrix test, JAIP matches the performance of CVM-JIT when the thread number is equal to 2 and outperforms CVM-JIT when the thread number is larger than 2. Finally, for the N-Queens test result shown in Fig. 17, JAIP outperforms CVM-JIT even if there is only one thread. This is probably because the NQueens program leaves little room for bytecode optimization by the JIT technique.

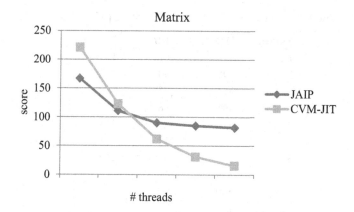

Fig. 16. JemBench scores of the 'Matrix' test on a single JAIP core.

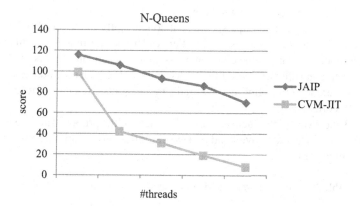

Fig. 17. JemBench scores of the 'N-Queens' test on a single JAIP core.

4.2 Multi-core Multithread Performance Evaluation

For multi-core multi-thread performance evaluation, we do not use CVM-JIT as a comparison point because the software platform does not support multi-core execution of Java applications. Here, we focus on the evaluation of performance scalability of JAIP-MP when the threads are distributed over multiple processor cores. As Table 2 shows, when the total number of threads is less than or equal to four, the JemBench score scales up fairly well (up to 3.69 times faster for the N-Queens test). When the total number of threads is more than four, the score of each benchmark naturally drops as the preemptive multi-threading mechanism of each JAIP core kicks in and there are synchronization overheads due to the way the benchmarks are designed. This is especially true for the N-Queens test.

Table 2. The multi-core JemBench scores of the parallel benchmarks.

# threads	Dummy	Matrix	N-Queens
1	151	167	116
2	298	240	225
3	374	395	330
4	491	498	428
5	410	425	311
6	373	412	251
8	362	399	262
12	359	366	212
16	340	327	195

Note: Larger number means better scores.

4.3 Synchronization Overhead

In the JemBench tests, context-switching overhead is not the only reason to cause the performance drop. If several requests are sent concurrently to the DCC of JAIP-MP, it takes several cycles for the mutex controller to decode the requests sequentially. In addition, to maintain data cache coherency, as soon as each entry is updated in any of the object heap controllers, the modified entry and its corresponding address are sent to the cache coherence controller and the main memory controller for cache validation among JAIP cores.

Tables 3 and 4 show the synchronization overhead of the proposed architecture under the Matrix test. The average overhead of a synchronization operation can be as small as tens of machine cycles.

Table 3. The execution time (in clock cycles) of acquiring a lock object.

# threads	4	6	8	12	16
Average	23.1	25.4	25.4	28.9	28.9
Worst-case	43	89	108	108	110
Best-case	9	9	9	9	9

Table 4. The execution time (in clock cycles) of releasing a lock object.

# threads	4	6	8	12	16
Average	20.2	21.6	21.8	28.4	28.7
Worst-case	43	76	85	92	98
Best-case	10	10	10	10	10

5 Conclusions and Future Work

In this chapter, we have presented an four-core Java processor, JAIP-MP. The uniqueness of JAIP-MP is that the key functions of an operating system kernel are implemented in hardware circuits. For thread management, the architecture supports arbitrary number of threads (limited by the on-chip TCB memory size), low context-switching overhead, small time quantum, and low synchronization overhead. The proposed architecture is implemented and verified on an FPGA platform. Experimental results show that the proposed design is very promising for embedded multi-thread applications.

For future work, we will look into the following directions. First of all, although the ping-pong buffer for context-switching is efficient performance-wise, it does impose heavy memory accesses. This may result in high power consumption. In the future, we will try to design a new architecture that can reduce the number of memory access per context-switch. Secondly, the coherent data cache in our current implementation only adopts one-level of cache hierarchy. Most general purpose processors nowadays adopt two or even three levels of cache hierarchy. It would be interesting to study the effects of a multi-level cache on the object-oriented programming model of the Java language.

Finally, current thread management design only uses a round-robin policy to maintain load balance. We will look into the design of a new architecture that can customize the thread distribution policy at runtime and allow for thread migration across different JAIP cores so that better runtime load balance can be achieved.

References

1. Ritchie, S.: Systems programming in Java. IEEE Micro **17**(3), 30–35 (1997)
2. Montague, B.R.: JN: OS for an embedded Java network computer. IEEE Micro **17**(3), 54–60 (1997)
3. Tsai, C.-J., Kuo, H.-W., Lin, Z., Guo, Z.-J., Wang, J.-F.: A Java processor IP design for embedded SoC. ACM Trans. Embed. Comput. Syst. **14**(2), Article 35 (2015)
4. Su, H.-C., Wu, T.-H., Tsai, C.-J.: Temporal multithreading architecture design for a Java processor. In: Proceedings of the IEEE International Symposium on Circuit and Systems (ISCAS 2014), Melbourne, Australia, June 2014
5. Schoeberl, M.: A Java processor architecture for embedded real-time systems. EUROMICRO J. Syst. Architect. **54**(1–2), 265–286 (2008)
6. Gruian, F., Schoeberl, M.: Hardware support for CSP on a Java chip multiprocessor. Microprocess. Microsyst. **37**(4), 472–481 (2013)

7. Brandner, F., Thorn, T., Schoeberl, M.: Embedded JIT compilation with CACAO on YARI. In: Proceedings of the IEEE International Symposium on Object/Component/Service-Oriented Real-Time Distributed Computing (ISORC 2009), Tokyo, 17–20 March, pp. 63–70 (2009)

8. Tyystjaervi, J., Saentti, T., Plosila, J.: Efficient bytecode optimizations for a multicore Java co-processor system. In: Proceedings of the 12th Biennial Baltic Electronics Conference, Tallinn, Estonia, 4–6 October 2010

9. Kreuzinger, J., Brinkschulte, U., Pfeffer, M., Uhrig, S., Ungerer, T.: Real-time event-handling and scheduling on a multithreaded Java microcontroller. Microprocess. Microsyst. 27(1), 19–31 (2003)

10. Sun Microsystems: picoJava-II Microarchitecture Guide (1999)

11. Uhrig, S., Wiese, J.: jamuth: an IP processor core for embedded Java real-time systems. In: Proceedings of the 5th International Workshop on Java Technologies for Real-Time and Embedded Systems (JTRES 2007), 26–28 September, Vienna, pp. 230–237 (2007)

12. Pitter, C., Schoeberl, M.: Towards a Java multiprocessor. In: Proceedings of the 5th ACM International Workshop on Java Technologies for Real-Time and Embedded Systems (JTRES 2007), 26–28 September, Vienna, pp. 144–151 (2007)

13. Yan, L., Liang, Z.: An accelerator design for speedup of Java execution in consumer mobile devices. Comput. Electr. Eng. 35(6), 904–919 (2009)

14. Hardin, D.S.: Real-time objects on the bare metal: an efficient hardware realization of the JavaTM virtual machine. In: Proceedings of the 4th IEEE International Symposium on Object-Oriented Real-Time Distributed Computing, 2–4 May, Magdeburg, pp. 53–59 (2001)

15. Vijaykrishnan, N., Ranganathan, N., Gadekarla, R.: Object-oriented architectural support for a Java processor. In: Proceedings of the 12th European Conference on Object-Oriented Programming, Brussels, Belgium, pp. 330–354 (1998)

16. Lin, Z.-G., Kuo, H.-W., Guo, Z.-J., Tsai, C.-J.: Stack memory design for a low-cost instruction folding Java processor. In: Proceedings of the IEEE International Symposium on Circuit and Systems, 20–23 May, Soeul, Korea, pp. 3326–3229 (2012)

17. Lindholm, T., Yelling, F.: The Java Virtual Machine Specification, 2nd edn. Addison-Wesley, Longman Publishing Co., Inc., Boston (1999)

18. Chang, Y., Wellings, A.: Garbage collection for flexible hard real-time systems. IEEE Trans. Comput. 59(8), 1063–1075 (2010)

19. Dijkstra, E.W., Lamport, L., Martin, A.J., Scholten, C.S., Steffens, E.F.M.: On-the-fly garbage collection: an exercise in cooperation. Commun. ACM 21(11), 965–975 (1978)

20. Bacon, D.F., Cheng, P., Shukla, S.: And then there were none: a stall-free real-time garbage collector for reconfigurable hardware. Commun. ACM 56(12), 101–109 (2013)

21. Srisa-an, W., Lo, C.-T.D., Chang, J.M.: Active memory processor: a hardware garbage collector for real-time Java embedded devices. IEEE Trans. Mobile Comput. 2(2), 89–101 (2003)

22. Gruian, F., Salcic, Z.A.: Designing a concurrent hardware garbage collector for small embedded systems. In: Srikanthan, T., Xue, J., Chang, C.-H. (eds.) ACSAC 2005. LNCS, vol. 3740, pp. 281–294. Springer, Heidelberg (2005)

23. Schoeberl, M., Preusser, T. B., Uhrig, S.: The embedded Java benchmark suite JemBench. In: Proceedings of the JTRES 2010, 19–21 August, Prague, Czech Republic (2010)

24. Oracle: Phoneme project webpage. Accessed 27 Sept 2011. https://java.net/projects/phoneme

Automatic Generation and Qualification of Assertions on Control Signals: A Time Window-Based Approach

Alessandro Danese$^{(\boxtimes)}$, Francesca Filini, Tara Ghasempouri,
and Graziano Pravadelli$^{(\boxtimes)}$

Department of Computer Science, University of Verona, Verona, Italy
{alessandro.danese,tara.ghasempouri,graziano.pravadelli}@univr.it,
francesca.filini@studenti.univr.it

Abstract. Assertion-based verification (ABV) is a promising approach for proving that the design implementation is consistent with the designer's intents. However, ABV effectiveness depends on the quality of the assertions that are defined to capture the designer's intents. Assertions are generally defined by verification engineers that manually convert informal specifications in logic formulas according to their expertise. However, manual definition is a time-consuming and error-prone activity, which may fail to exhaustively cover either the intended specification or the implemented behaviours. For this reason, different mining approaches have been recently proposed for the automatic generation of assertions. Unfortunately, in most cases, existing mining tools generate a set of over-constrained assertions. As a consequence, each assertion in the set is a long formula that describes a very specific behaviour of the design under verification (DUV). Thus, in the effort of covering as much DUV behaviours as possible, these approaches generate a huge amount of assertions with a negative impact on the total time required for their verification. To overcome this drawback, this paper introduces a dynamic approach that incrementally analyses control signals on DUV execution traces for mining more expressive temporal assertions that better capture the I/O communication protocol. Then, to evaluate the effectiveness of the generated assertions in covering the intended behaviours, a technique is proposed to estimate assertion's interestingness by ranking them according to metrics typically adopted in the context of data mining.

Keywords: Assertion mining · Assertion qualification · Assertion-based verification

1 Introduction

Despite the advancement in simulators and formal methods, the verification result is only as good as the specifications defined to capture the designer's

© IFIP International Federation for Information Processing 2016
Published by Springer International Publishing AG 2016. All Rights Reserved
Y. Shin et al. (Eds.): VLSI-SoC 2015, IFIP AICT 483, pp. 193–221, 2016.
DOI: 10.1007/978-3-319-46097-0_10

intent. Without a good set of specifications, designers can easily lose the control of the system design, and they cannot understand what behaviour of the system is implemented in the reality. Unfortunately, even if everybody agrees that the first and the most important step for a good development task is the collection of all the intents and specifications, often this does not receive an appropriate attention. In the last decade, assertion-based verification (ABV) has arisen as one of the most popular solutions for electronic system level (ESL) and Register Transfer Level (RTL) verification [1]. ABV relies on the definition of assertions, i.e., logic formulas, generally written according to temporal logics, like LTL and CTL [2], and property specification languages, like PSL [3], that formalize the behaviours of the DUV by overcoming the ambiguity of natural languages and providing the engineers with precise and well-defined specifications. However, manual definition of assertions requires high expertise and it is an error-prone and time-consuming activity. Main problems are related to the risk of defining assertion sets that are incomplete (i.e., unable to cover all expected behaviours of the DUV), inconsistent (i.e., with contradicting assertions), redundant (i.e., with assertions that are logical consequence of others), and including vacuous assertions (i.e., assertions that are true independently from the DUV, and thus irrelevant). As a result, a false sense of security is induced by an ABV campaign conducted with a low-quality set of assertions. As a complementary approach to manual definition of assertions, several approaches and tools have been proposed for automatically extracting safety assertions in the form $always(antecedent \rightarrow consequent)$ from the implementation of the DUV [4–8]. These approaches either rely on static analysis of the DUV source code or they dynamically mine assertions from execution traces of the DUV. The first are accurate and provide assertions that are formally proved to be satisfied by the DUV, but they do not scale well for complex DUVs. The second provide only likely assertions, whose quality depends on the observed execution traces (i.e., likely assertions are guarantee to hold at least for the considered execution traces, however a counter example could be finally found), but they do not require the source code and guarantee a better scalability for complex DUVs. Independently from the adopted techniques, mined assertions can be compared with design intents to discover unexpected behaviours implemented in the DUV, to confirm that relevant behaviours are actually implemented, and for documentation purposes. Existing approaches generally extract a high number of long assertions, but each of them covers a few specific behaviours of the DUV. This depends on the fact that mined assertions are over-specified, i.e., their antecedents and/or consequents include several atomic propositions that predicate almost on all primary inputs and outputs of the DUV. Moreover, mined assertions mix data and control signals making difficult the characterization of the I/O communication protocol of the DUV. Large sets of over-constrained assertions make impractical the analysis of mined assertions by verification engineers and greatly increase verification time. Furthermore, while vacuity and inconsistency in the set of generated assertions are generally avoided by the mining approach itself, assertion incompleteness and redundancy may still affect the outcome of assertion mining. Thus, a qualification phase to identify the most interesting assertions is necessary to focus designer's attention on relevant assertions that capture expected (or unexpected) behaviours implemented in the DUV. Unfortunately, as the number of mined assertions can be very high, their manual qualification is almost impractical, while current approaches for automatic evaluation are still unsatisfactory from the point of view either of the effectiveness [5,6,9,10] or of the efficiency [11–14].

In the previous context, the goal of this paper consists of overcoming the drawbacks of existing works in assertion generation and assertion qualification by proposing:

- First, a dynamic *assertion generation* approach that infers assertions by incrementally analysing time windows of the DUV's execution traces. The algorithm searches for recurrent temporal patterns among atomic propositions predicating on I/O control signals. Mined assertions are in average shorter, more expressive from the point of view of the I/O communication protocol, and then simpler to be understood by humans, compared, for example, to the approach proposed in [5]. This reduces verification time, and increases the effectiveness of verification engineers in discovering design errors by analysing mined assertions.
- Secondly, an automatic *qualification* approach for evaluating the quality of extracted assertions and rank them accordingly. The estimation of the interestingness of assertions is achieved by ranking them according to probabilistic metrics typically adopted in the context of data mining (i.e., support and correlation coefficient) [15,16], which we adapt here for the specific case of assertion mining. From the point of view of the general concept, data mining and assertion mining share the same idea (extracting rules from data), but they have several differences that make practically different how these metrics are computed and interpreted for evaluating the interestingness of assertions.

The rest of the paper is organized as follows. Section 2 summarizes the related works. Section 3 presents some preliminary definitions and concepts relevant for understanding the technical details of our approach. Section 4 provides an overview of the proposed methodology, which is then thoroughly described in Sects. 5 and 6, respectively, for assertion mining and assertion qualification. Finally, experimental results and concluding remarks are reported, respectively, in Sects. 7 and 8.

2   Related Works

Different strategies have been proposed for assertion mining. Among the first works in the software domain, scenario-based specification mining approaches proposed in [17,18] require instrumenting the source code to store the sequence of method calls among the components of the design in an execution trace. Data mining methods are then applied to program execution traces in order to mine strongly observed inter-component universal sequence diagrams in the form of live linear sequence charts (LSC). LSCs represent how the components cooperate to implement certain system features. However, these approaches are not aimed at discovering the complete behaviour of the components but only the collaboration among them. Other works mine the specifications of the design in form of algebraic equation [19] or Hoare-style equations of pre and post-conditions [20]. In particular, [20] automatically deducts arithmetic relationships that predicate exclusively on values and data-structures, but the temporal behaviours are not considered. Temporal assertion mining is described in [21], where a mining tool, GoldMine, is proposed for extracting Boolean-level assertions for HW components. On the contrary, [6,9,22] propose methodologies for specification mining which are able to manage indistinctly HW and embedded SW descriptions according to a set of pre-defined temporal patterns. They rely on Daikon [23] to mine relevant arithmetic/logic relations among the variables of the DUV from execution traces. In particular, [6,22] try to mine a set of temporal assertions reporting how arithmetic/logic relations change during the execution of the design. Even if these works introduce a novel approach in which the extraction of temporal relations between arithmetic/logic expressions represents a good strategy to describe more closely the behaviours of the design, the generated assertions are hard to be understood because they involve all primary inputs and outputs of the DUV. As results, mined assertions cannot be easily analysed from designers to detect unexpected behaviours. On the other hand, [9] is focused only on catching simple arithmetic/logic relations on the assumption that a behaviour of the DUV can be justified essentially through a comparison between the set of verified arithmetic/logic relations in an execution instant and the set of falsified arithmetic/logic relations in the closer next execution instants. Finally, commercial tools are also available for automatic assertion generation, e.g., Atrenta BugScope [24] and Jasper ActiveProp [7] that works on RTL models at Boolean level, but no arithmetic/logic expressions are considered among more abstracted data types.

Concerning assertion qualification, current approaches are still unsatisfactory to measure the quality and interestingness of assertions. In [6], a stressing phase is proposed only to verify the likelihood that mined assertions are globally satisfied (and not only for the execution traces analysed by the miner), but no strategy is proposed to measure their interestingness in covering DUV behaviours. In [5], interestingness estimation is based on the number of propositions included in the antecedent of the assertion, according to the fact that an assertion with a lower number of propositions in its antecedent has an higher input space coverage than one with many propositions in its antecedent. However, the correlation between the antecedent and the consequent of an assertion is not considered. To solve this drawback, in [9] a ranking function is proposed that evaluates the quality of the mined assertions in terms of cause-effect relationship between antecedent and consequent of an assertion. Finally, in [10], mined assertions are said to be generally ranked according to their frequency of occurrences and time of first occurrence but no specific approach is presented.

As an opposite class of approaches, coverage metrics have been widely studied for qualification of assertions [11–14]. Most of these works relies on mutation analysis, which requires perturbing the DUV implementation by injecting mutations (faults) to check, either statically [12,13] or dynamically [14], whether they change the truth values of the assertions; mutations that do not cause a change are said to be not detected. Assertions that detect a few mutations are less interesting than assertions detecting an higher number of mutations. Not detected mutants generally highlight area/behaviours of the DUV that are not covered by any of the defined assertions showing a hole on the coverage. Dynamic approaches like [14] scale better with respect to static techniques, however, they still require long simulation runs for checking each assertion for each mutation with a significant set of testbenches. When the number of assertions is very high, as in the case of assertions extracted automatically, evaluating their interestingness through mutation analysis becomes a very time-consuming activity.

3   Background and Preliminaries

Before describing the proposed methodology, some definitions and concepts concerning assertion generation and assertion qualification are reported to create the necessary background.

3.1   Definitions

Definition 1 (Execution trace).   *Given a finite sequence of simulation instants* $\langle t_1,...,t_n \rangle$ *and a model M working on a set of variables V, an execution trace of M is a finite sequence of pairs $T = \langle (V_1,t_1),...,(V_n,t_n) \rangle$ where $V_i = eval(V,t_i)$ is the evaluation of variables in V at simulation instant t_i.*

More informally, an execution trace describes for each simulation instant t_i the values assumed by each variable included in V during the evolution of the design

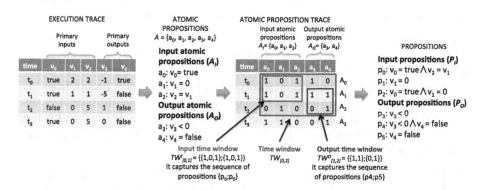

Fig. 1. Exemplification of execution trace, atomic proposition trace and time window. (Color figure online)

under verification M. In this paper, variables in V are primary inputs and primary outputs representing control signals of the DUV. An example of an execution trace is reported in Fig. 1 (left). By analysing an execution trace, we can extract a set of atomic propositions that predicate on variables included in V.

Definition 2 *(Atomic proposition). An atomic proposition is a formula that does not contain any logical connective.*

In our methodology, a set of atomic propositions is organized in an array $A = \{a_1, \ldots, a_m\}$ which is further divided in two sub-arrays: the array of *input atomic propositions* $A_I = \{a_0, \ldots, a_i\}$, whose elements are atomic propositions that predicate only on primary inputs of the DUV, and the array of *output atomic propositions* $A_O = \{a_{i+1}, \ldots, a_m\}$, whose elements predicate only on primary outputs of the DUV. We do not consider any atomic proposition that mixes primary inputs and primary outputs. Examples of atomic propositions are reported in Fig. 1 (centre). From an execution trace and the corresponding set of atomic propositions, we can generate an atomic proposition trace.

Definition 3 *(Atomic proposition trace). Given an execution trace T and an array of atomic propositions A, an atomic proposition trace is a finite sequence of pairs $\omega = \langle (A_0, t_0), \ldots, (A_n, t_n) \rangle$ where $A_i = eval(A, t_i)$ is an array that represents the evaluation of atomic propositions in A at simulation instant t_i, i.e., $A_i[j] = 0$ if $A[j] = false$ at time t_i, $A_i[j] = 1$ otherwise.*

The atomic proposition trace is the data structure we use to mine propositions.

Definition 4 *(Proposition). A proposition is a composition of atomic propositions through logic connectives. An atomic proposition itself is a proposition.*

It this paper, we consider propositions involving only the logic *and* ($\wedge$) as connective, and we classify them in two different sets:

- The set of *input propositions* (P_I): a proposition p belongs to P_I if it is composed only of input atomic propositions and $\wedge$ connectives;
- The set of *output propositions* (P_O): a proposition p belongs to P_O if it is composed only of output atomic propositions and $\wedge$ connectives.

An example of an atomic proposition trace and a set of input/output propositions that can be extracted from it are shown in Fig. 1 (right). To represent input and output propositions in a compact and efficient way, we use an array-based notation. Given an array of input (respectively, output) atomic propositions A_I (A_O), an input (output) proposition is represented by an array of Boolean values p such that $p[i] = 0$ if the input (output) atomic proposition $A_I[i]$ ($A_O[i]$) is not used in the proposition, $p[i] = 1$ otherwise. For example, the input proposition p_0 in Fig. 1 can be represented by the array $\{1, 0, 1\}$.

Definition 5 *(Time window). Given an atomic proposition trace $\omega = \langle (A_0, t_0), \ldots, (A_n, t_n) \rangle$, and two simulation instants t_i and t_j such that $1 \le t_i \le t_j \le n$, a time window $TW_{[i,j]} = \langle (A_i, t_i), \ldots, (A_j, t_j) \rangle$ is the subsequence of contiguous elements of ω included between instant t_i and instant t_j.*

Given a time window $TW_{[i,j]}$, and a simulation instant t_k such that $t_i \leq t_k \leq t_j$, we can separate $TW_{[i,j]}$ in two parts:

- The *input time window* $TW_{[i,k]}^I$, which is composed of elements of $TW_{[i,j]}$ included in the simulation instants between t_i and t_k, restricted to the input atomic propositions.
- The *ouptut time window* $TW_{[k,j]}^O$, which is composed of elements of $TW_{[i,j]}$ included in the simulation instants between t_k and t_j, restricted to the output atomic propositions.

The input time window $TW_{[i,k]}^I$ and the corresponding output time window $TW_{[k,j]}^O$ overlap for exactly one simulation instant (t_k). Given a time window $TW_{[i,j]}$, we can generate $j - i + 1$ different couples of input/output time windows, one for each simulation instant $t_k \in [t_i, t_j]$. For example, in the atomic proposition trace of Fig. 1, the green box highlights a time window composed of 3 simulation instants in the interval $[t_0, t_2]$, the red box corresponds to an input time window in the interval $[t_0, t_1]$, and finally the blue box shows the corresponding output time window in $[t_1, t_2]$. In the rest of the paper we will represent an input (output) time window by the sequence of arrays corresponding to the input (output) propositions it captures. For example, in Fig. 1, the input time window corresponding to the interval $[t_0, t_1]$ is represented by the sequence of two arrays $\{\{1, 0, 1\}; \{1, 0, 1\}\}$.

By analysing input and output time windows in the atomic proposition trace we can mine temporal assertions.

Definition 6 (Temporal assertion). *A temporal assertion is a composition of propositions through temporal operators and logic connectives.*

In this paper, we consider Linear Temporal Logic (LTL) assertions in the form $G(antecedent \rightarrow consequent)$, where G is the LTL *always* operator[1], and *antecedent* and *consequent* may involve only X, i.e., the LTL *next* operator[2] and the $\wedge$ connective. Moreover, *antecedent* is composed only of an arbitrary number of propositions belonging to the set P_I extracted by analysing an input time window $TW_{[i,k]}^I$, while *consequent* includes a single proposition belonging to the set P_O extracted by analysing the corresponding output time window $TW_{[k,j]}^O$. We selected this specific form of assertion, since it is suited to describe the behaviour of the I/O communication protocol of a DUV, which, as reported in the introduction, represents the target for the current work. For example, from the atomic proposition trace of Fig. 1 the following temporal assertion can be mined: $G((p_0 \wedge X^2(p_1)) \rightarrow X^3(p_5))$.

In according to the above definitions and notations, we define hereafter a set of operators working on propositions that will be used to illustrate the mining methodology proposed in Sect. 5.

---

[1] Given a formula α, $G(\alpha)$ means that α is always true.
[2] Given a formula α, $X(\alpha)$ means that α is true at the next instant. As a short cut, we will write $X^n(\alpha)$ to represent the application of n consecutive next operators to the formula α.

RemoveAtomicProp: Given an array of Boolean elements p corresponding to a proposition, and an index i corresponding to the atomic proposition a_i, which represents the element at position i of p, $RemoveAtomicProp(p, i)$ returns a new array p' where $p'[j] = p[j]$ for all indexes $j \neq i$ and $p'[i] = 0$, i.e., the proposition corresponding to p' does not include a_i.

AddAtomicProp: Given an array of Boolean elements p corresponding to a proposition, and an index i corresponding to the atomic proposition a_i, which represents the element at position i of p, $AddAtomicProp(p, i)$ returns a new array p' where $p'[j] = p[j]$ for all indexes $j \neq i$ and $p'[i] = 1$, i.e., the proposition corresponding to p' includes a_i.

Overlapping: Given two arrays of Boolean elements p and q corresponding to two propositions predicating over the same array of atomic propositions, $Overlapping(p, q)$ returns $true$ if $q[i] = 1$ for at least all indexes i such that $p[i] = 1$, it returns $false$ otherwise.

Complement: Given two arrays of Boolean elements p and q corresponding to two propositions predicating over the same array of atomic propositions, $Complement(p, q)$ returns a new array p' where $p'[i] = false$ for all indexes i such that $q[i] = false$, and $p'[j] = p[j]$ for all indexes j such that $q[i] = true$.

For example, Fig. 2 shows the application of operators $RemoveAtomicProp$, $Complement$ and $Overlapping$ to propositions. In a more general way, we apply $Complement$ and $Overlapping$ also to sequences of propositions of the same length. In this case, given two sequences of propositions s_1 and s_2, $Complement$ and $Overlapping$ operate iteratively on each couple of corresponding elements (p_i, q_i) such that $p_i \in s_1$ and $q_i \in s_2$.

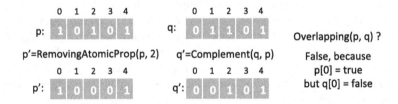

Fig. 2. Examples of application of $RemoveAtomicProp$, $Complement$ and $Overlapping$.

In addition to the previous definitions, which are necessary to understand the assertion mining approach detailed in Sect. 5, the rest of this section summarizes some concepts concerning data mining, such as itemsets, transactions and association rules, which are necessary to present the assertion qualification methodology proposed in Sect. 6.

Definition 7. *Let $I = \{i_1, i_2, \ldots, i_n\}$ be a set of items. Let $D = \{d_1, d_2, \ldots, d_m\}$ be a data set, i.e., a set of observations, called transactions, with respect the set of items I. Each element in D contains a subset of the items in I. An association* *rule is defined as an implication of the form* $X \rightarrow Y$ *where* $X, Y \subseteq I$ *and* $X \cap Y = \emptyset$. X *and* Y *are called itemsets.*

Figure 3(a) shows an example of a data set which describes the behaviours of customers in a supermarket with respect to a set of items (i.e., milk, bread, ..., coffee). Data mining approaches are generally intended to extract association rules from data sets, which are then used to predict non trivial, implicit, previously unknown and potential useful information, like, for example, "when milk is bought bread and coffee are generally bought too", which is expressed by the association rule $Milk \rightarrow Bread \wedge Coffee$. Assertion mining deals instead with executing traces and temporal assertion which are formalized in Definitions 1 and 6 respectively. An example of a temporal assertion in Linear Time Logic is $always(p_1 \rightarrow next(p_2 \wedge p_3))$ which states it always happens that p_2 and p_3 are satisfied one simulation instant later than p_1 becomes true (Fig. 3(b)).

In this paper, without lack of generality, and to preserve the independence of the proposed qualification methodology from specific temporal patterns instantiated in the analysed assertions, we generically consider formulas in the form $A \rightarrow C$, where the antecedent A and the consequent C are composed by propositions, logic connectives, and temporal operators according to the selected temporal logic. The initial hypothesis is that the analysed assertions are true on the DUV.

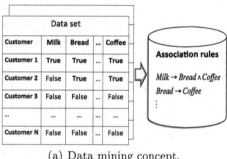

(a) Data mining concept.

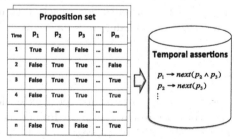

(b) Assertion mining concept.

Fig. 3. Similarities between data mining (a) and assertion mining (b).

3.2   Comparing Data Mining and Assertion Mining

The overall goal of data mining is to extract information from a data set and transform it into an understandable and useful structure. This structure allows user analysing data from many different dimensions, categorizing them and summarizing correlations between items in a database. For example, analysing data from behaviours of different customers as reported in Fig. 3(a) leads to obtain useful information and helps analysers to decide which trend is more interesting for marketing. Association rules can also be extracted when data are referred to time sequences. In this case, temporal data mining strategies are adopted, whose goal is to discover hidden relations between sequences and sub sequences of events [25]. In any case, the mined (temporal) association rules are a prediction for future behaviours, which may be true or not. Metrics are thus used to estimate the probability that rules extracted from past observations can be valid also in the future. On the contrary, the main goal of assertion mining consists of extracting formulas that exactly describe the functionality implemented in the DUV, which is not ambiguous and does not vary in the future, except in the case the implementation is changed. Assertion mining is thus not intended to predict the future, but to formalize the actual set of DUV behaviours. Summarizing, main similarities among data mining and assertion mining are the presence of a set of data that represents observations with respect to past behaviours exposed by the observed target (customers, DUV, ...), and the need of extracting association rules that formalize such observations. As shown in Fig. 3(a) and (b), items, data sets, and association rules in data mining correspond, respectively, to propositions, execution traces, and temporal assertions in assertion mining. Meanwhile, the main difference between data mining and assertion mining is represented by the concept of transaction (i.e., a row in a data set), which does not have a direct correspondence with a row of an execution trace, because a temporal assertion is composed by one antecedent and one consequent that are true in different instants inside the execution trace. This difference impacts on the way metrics typically adopted for evaluating association rules in data mining can be reused for measuring the interestingness of assertions. Finally, another difference is related to the final goal of the mining: in one case the prediction of future behaviours, in the other the formalisation of actual (unmodifiable, except in the case the DUV functionality is changed) behaviours.

4   Methodology

The proposed methodology (Fig. 4) starts from a set of execution traces obtained by simulating the DUV, which represent the input for the assertion miner. The miner first extracts atomic propositions representing the building blocks and then compose them to generate temporal assertions. Finally, the generate assertions are evaluated according to their degree of interestingness and a final ranked set of assertions is provided.

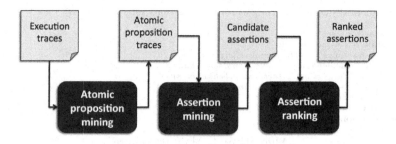

Fig. 4. Methodology overview.

5   Assertion Mining

Given a DUV and a related set of execution traces, the proposed methodology consists of two main phases: (i) mining of atomic propositions and generation of the corresponding atomic proposition traces, and (ii) extraction of input and output time windows and mining of temporal assertions. Since, the first phase has been already faced in [6], this paper focuses only on the second phase.

The main procedure of the proposed approach is showed in Algorithm 1. The function MAIN takes as parameters an atomic proposition trace w and two thresholds: tw_len, and max_len, with $max\_len \leq tw\_len$. The first threshold (tw_len) advises how many clock cycles are required, at maximum, by the DUV to compute its functionality. This parameter, which can generally be obtained from the documentation or from simulation of the DUV, represents the length of the time windows that will be analysed by the miner inside w. The second threshold (max_len) is used to bound the maximum number of instants that will be considered to mine the antecedent of a temporal assertion, i.e., it represents the maximum length for an input time window. Several DUVs need a high number of clock cycles for completely computing their functionality, but only few of them could be necessary to set control signals such that the computation phase starts. For instance, to mine interesting antecedents for an encryption/decryption component, it is relevant catching what happens to control signals during the initialization phase, before the computation proceeds without affecting control signals till outputs are ready. Similarly, in the case of a network component, it is relevant to capture in the antecedent of an assertion how control signals change during the phase that makes data transmission starting. The max_len threshold is then intended to guide the mining procedure such that, for mining antecedents, it focusses on the initial part of the considered time windows when the DUV is characterized, after an initialization phase, by a long elaboration phase that does not involve control signals any more till results are ready. In this case, generally max_len is small with respect to tw_len. On the contrary, max_len is set to be equal to tw_len for DUVs where control signals can change the functionality of the DUV at each simulation instant.

```
 1. function MAIN(ω, tw_len, max_len) return assertions_set
 2.     len = 1
 3.     assertions_set = ∅
 4.     while len ≤ max_len do
 5.         behaviours = getBehaviours(ω, tw_len, len)
 6.         pruningBehaviours(assertions_set, behaviours)
 7.         candidates = getAssertions(ω, behaviours)
 8.         assertions_set = getAssertions(ω, behaviours)
 9.         len = len + 1
10.     end while
11. end function
```

Algorithm 1. Main procedure.

After initializing the *assertions_set* to the empty set, the main loop of Algorithm 1 cyclically calls the following three functions, by varying the length (*len*) of the considered input time window at each iteration, while $len \leq max\_len$.

1. *getBehaviours*(ω, tw_len, len): it detects the behaviours exposed by the DUV in the atomic proposition trace. Each behaviour is represented by a sequence (with length between 1 and *len*) of input propositions associated to a unique output proposition. For example, in the atomic proposition trace of Fig. 1, each time the input proposition $p_0 = \{1, 0, 1\}$ (corresponding to $a_0 \wedge a_2$) is true at a simulation instant t_i, the output proposition $p_5 = \{0, 1\}$ (corresponding to a_4) holds at simulation instant t_{i+1}. Thus, a behaviour can be identified that associates the input proposition p_0 to the output proposition p_5 with an *offset* of one simulation instant.
   The second behaviour associates p_0 to the output proposition $p_5 = \{0, 1\}$, because each time p_0 holds, the output proposition a_4 is satisfied in the next simulation instant.

2. *pruningBehaviours*(assertion_set, behaviours): it removes all the behaviours returned by *getBehaviours* that are already captured by assertions mined in the previous iterations of the main procedure. For example, if the assertion $\alpha = G(a_2 \rightarrow X(a_4))$ is already included in the *assertion_set*, a behaviour, which associates, with an offset of one simulation instant, the input proposition p_0 to the output proposition p_5, can be discarded, since it will create the assertion $G(a_0 \wedge a_2 \rightarrow X(a_4))$ that is a logical consequence of α.

3. *getAssertions*(ω, behaviours): it is in charge of creating a set of candidate temporal assertions, compliant with the form described in Subsect. 3, starting from the behaviours survived after *pruningBehaviour* is called.

The next sections describe in details how the previous functions work to mine temporal assertions.

5.1   Mining of Interesting Behaviors

The function *getBehaviours* is implemented as shown in Algorithm 2. It takes as arguments an atomic proposition trace ω and the two thresholds *tw_len* and *len*. Its goal is to detect behaviours exposed in the execution traces of the DUV,

```
 1. function GETBEHAVIOURS(ω, tw_len, len) return behaviours
 2.     behaviours = ∅
 3.     dictionary = ∅
 4.     t_i = 0
 5.     while t_i ≤ (length(ω) − tw_len) do
 6.         in = propSeq(TW^I_[t_i,t_i+len−1])
 7.         out = propSeq(TW^O_[t_i+len−1,t_i+tw_len−len+1])
 8.         ⟨in, old_out⟩ = findInDictionary(dictionary, in)
 9.         if ⟨in, old_out⟩ = NIL then
10.             dictionary = dictionary ∪ ⟨in, out⟩
11.         else
12.             new_out = Complement(old_out, out)
13.             dictionary = dictionary \ ⟨in, old_out⟩
14.             dictionary = dictionary ∪ ⟨in, new_out⟩
15.         end if
16.         t_i = t_i + 1
17.     end while
18.     for all ⟨in, out⟩ ∈ dictionary do
19.         offset = 0
20.         for all p ∈ out do
21.             if p ≠ NIL then
22.                 behaviours = behaviours ∪ ⟨in, offset, p⟩
23.             end if
24.             offset = offset + 1
25.         end for
26.     end for
27. end function
```

Algorithm 2. Extraction of DUV behaviours.

in the form of associations between a sequence of input propositions and a corresponding sequence of output propositions. In order to detect such associations, a time window $TW_{[t_i,t_i+tw\_len−1]}$ of length tw_len is analysed for each simulation instants t_i (lines 5–17). Such a time window is decomposed in an input time window of length len, and a corresponding output time window of length $tw\_len − len + 1$. Then, from the input time window, the function *propSeq* extracts the corresponding sequence of input propositions (in) that hold in the interval $[t_i, t_i + len − 1]$ (line 6). Similarly, *propSeq* extracts from the output time window the sequence of output propositions (out) that hold in the interval $[t_i + len − 1, t_i + tw\_len − len + 1]$ (line 7). For example, let us consider the atomic proposition trace of Fig. 1 and let we fix $tw\_len = 2$ and $len = 1$. At time t_0, we have $in = \{1, 0, 1\}$ while out is represented by the sequence $\{\{1, 0\}; \{1, 1\}\}$. The function then searches if in is already present in the dictionary of the collected input/output propositions pairs (line 8–9). If this is not the case (line 10), a new association $\langle in, out \rangle$ is added to the dictionary. On the contrary, when in is already present, the operator *Complement* is applied to return a new sequence of output propositions new_out, where atomic propositions excluded from elements of out are excluded also from the corresponding elements of new_out. Then, the couple $\langle in, old\_out \rangle$ is replaced in the dictionary by $\langle in, new\_out \rangle$ (lines 12–14). This replacement is necessary to refine the already collected behaviours by removing from their output propositions the atomic propositions that become false in the current output time window, such that only behaviours that are never violated throughout the trace are finally collected. The replacement happens, for example, at simulation instant t_1 of the atomic proposition trace of Fig. 1. In fact, at t_1, $in = \{1,0,1\}$ is associated to $out = \{\{1,1\};\{0,1\}\}$. However, in was already associated to $old\_out = \{\{1,0\};\{1,1\}\}$ at t_0. Thus, in the dictionary $\langle\{1,0,1\},\{\{1,0\};\{1,1\}\}\rangle$ is replaced by $\langle\{1,0,1\},\{\{1,0\};\{0,1\}\}\rangle$ after the application of $Complement(old\_out, out)$. After the creation of the dictionary that collects associations between sequences of input and output propositions, the final loop (lines 18–26) creates a set of behaviours for each pair $\langle in, out\rangle$. In particular, a behaviour, represented by a triplet $\langle in, offset, p\rangle$, is extracted for each proposition p captured in the sequence of output propositions out, where $offset$ represents the distance, computed in simulation instants, between the last element of in and p. For example, for the pair $\langle\{1,0,1\},\{\{1,0\};\{0,1\}\}\rangle$ the following two behaviours are extracted: $\langle\{1,0,1\},0,\{1,0\}\rangle$ and $\langle\{1,0,1\},1,\{0,1\}\rangle$ to represent respectively that the output proposition $\{1,0\}$ holds exactly at the same time of input proposition $\langle\{1,0,1\}$ ($offset$ is 0), while the output proposition $\{0,1\}$ holds one simulation instant later ($offset$ is 1).

5.2   Pruning of Behaviours

The function $pruningBehaviours$ takes as arguments a set of assertions and a set of triplets representing behaviours collected by $getBehaviours$ in the form $\langle in, offset, p\rangle$. The goal of $pruningBehaviours$ is to preserve only the triplets that are not already covered by assertions collected in previous iteration of Algorithm 1. A triplet is covered by an assertion α when the following conditions are true concurrently:

1. All input atomic propositions included in the antecedent of α are also present in in (i.e., values assigned to primary inputs of the DUV that satisfy the atomic propositions included in in satisfy also the antecedent of α).
2. All output atomic propositions included in the consequent of α are also present in p (i.e. the consequent of α is at least as detailed as p).
3. The distance, computed in simulation instants, between the last input atomic proposition of the antecedent of α and the consequent of α equals $offset$.

Triplets that falsify at least one of the previous conditions are preserve, the others are discarded.

5.3   Mining of Assertions

Given an atomic proposition trace ω and a set of triplets of the form $\langle in, offset, p\rangle$ representing behaviours preserved by the $pruningBehaviours$ function, the $getAssertions$ function works as described in Algorithm 3. Its goal is to extract an assertion of the form $G(antecedent \rightarrow consequent)$ from every triplet, such that the input propositions $\{p_0, \ldots, p_i\}$ captured inside the sequence in act as an antecedent of the form $(p_0 \wedge \cdots \wedge X^i(p_i))$, while $X^{i+offset}(p)$ represents the consequent. This is performed by the $makeAss$ function at line 5. After an assertion is added to the set of candidates ass_set, the $pruningBehaviours$

```
1.  function GETASSERTIONS(ω, behaviours, len) return ass_set
2.      ass_set = ∅
3.      for all ⟨in, offset, p⟩ ∈ behaviours do
4.          s_in = simplify(⟨in, offset, p⟩)
5.          ass_set = ass_set ∪ makeAss(s_in, offset, p)
6.          pruningBehaviours(ass_set, behaviours)
7.      end for
8.  end function
9.
10. function SIMPLIFY(ω, ⟨in, offset, p⟩)
11.     len = length(in)
12.     p_off = len − 1 + offset
13.     for all q ∈ in do
14.         s_q = q
15.         i = 0
16.         while i < length(q) do
17.             s_q = RemoveAtomicProp(s_q, i)
18.             new_in = (in \ q) ∪ s_q
19.             t_i = 0
20.             while t_i ≤ (length(ω) − p_offset) do
21.                 temp = propSeq(TW^I_{[t_i, t_i+len−1]})
22.                 if Overlapping(new_in, temp) then
23.                     temp = propSeq(TW^O_{[p_off, p_off]})
24.                     if !Overlapping(p, temp) then
25.                         s_q = AddAtomicProp(s_q, i)
26.                         break
27.                     end if
28.                 end if
29.                 t_i = t_i + 1
30.             end while
31.         end while
32.         in = (in \ q) ∪ s_q
33.     end for
34.     return in
35. end function
```

Algorithm 3. Generation of temporal assertions.

function is called to remove behaviours implicitly covered by the new assertion (line 6).

In order to increase the DUV behaviours covered by the mined assertions, before calling *makeAss*, the input proposition included in the sequence *in* is first simplified by removing atomic propositions from the antecedent such that the consequent can be verified by a higher number of simulation instants, thus enforcing the final assertion (line 4). For instant, let us consider the triplet $\langle in, offset, p \rangle$ represented by $\langle \{\{1, 1, 0\}\}, 0, \{\{0, 1\}\} \rangle$. Looking at Fig. 1, we can see that the output proposition $\{0, 1\}$ is true at both simulation instants t_2 and t_3, but the input proposition $\{1, 1, 0\}$ is verified only at simulation instant t_3. However, if we set to *false* the first input atomic proposition of $\{1, 1, 0\}$, we obtain the proposition $\{0, 1, 0\}$, which is true at both simulation instant t_2 and t_3. Thus, $in = \{\{1, 1, 0\}\}$ can be replaced by $s\_in = \{\{0, 1, 0\}\}$ in the triplet to cover a wider time window in the atomic proposition trace. In this way, the assertion $G(a_1 \rightarrow a_4)$ can be extracted instead of $G(a_0 \wedge a_1 \rightarrow a_4)$. The first is preferred because it implies the second. The simplification of the sequence of input propositions *in* is performed by the function *SIMPLIFY* (lines 10–33). Given an input propositions q belonging to the sequence *in*, it makes a copy s_q

of q (line 14), and then it performs the following steps for each atomic proposition a_i included in s_q (lines 16–31):

- remove a_i from s_q (line 17);
- create a new sequence of propositions new_in from in by replacing the proposition q with s_q (line-18);
- check, for every simulation instant t_i, if the new sequence new_in is true on the atomic proposition trace (line-22), but the output proposition p is false (line 24). In this case, a counter example is found that shows we cannot remove the atomic proposition a_i from s_q, otherwise the association between s_q and p is not valid any more. Thus a_i is restored inside s_q (line 25). If a counter example is not found, a_i can be definitely removed.

6   Assertion Qualification

The degree of interestingness of assertions extracted by applying the methodology described in the previous section is evaluated according to a re-adaptation of metrics that are traditionally adopted in the context of data mining. Assertions are then ranked according to such a metrics.

6.1   Metrics

Several metrics have been proposed in data mining for evaluating the interestingness of association rules. The use of metrics allows analysers evaluating the rules from different points of view [15,26]. For instance, *odds ratio* and *entropy* are appropriate for estimating the probability of distribution of items, *support* and *confidence* are able to calculate the interestingness of an association rule based on the number of item's occurrences; while the *correlation coefficient* is suited to determine the dependency between set of items.

In the context of assertion qualification, metrics that provide information about the degree of accuracy of a rules with respect to the probability it will hold in the future (like for example, confidence, which estimates the joint probability between occurrences of the antecedent and the consequent in the data set) are not relevant, because we know that assertions under analysis are always true on the DUV. We are instead interested in metrics that measure the interestingness of an assertion with respect to covered behaviours, number of activations, and correlation between antecedents and consequents. For this reason, we identified *support* and *correlation coefficient* as the most interesting metrics for assertion evaluation. Their definition in the context of data mining are hereafter reported together with considerations related to how they can be adapted to be suited for assertion evaluation.

Definition 8. *Given a set of items I, and the corresponding set of transactions D, a rule $X \rightarrow Y$ has support S if X and Y occur concurrently in S percent of transactions in D.*

In practice, to compute the support of an association rule, it is necessary to count how many rows in the transaction set table contain both X and Y. In case of temporal assertions, the support corresponds instead to the number of times a temporal assertion occurs (i.e., its antecedent is fired and then its consequent is satisfied) in the execution traces with respect to the total number of occurrences corresponding to the other temporal assertions under analysis. This requires a different computation approach with respect to data mining. For example, let us consider a temporal assertion $A \to C$ that occurs 10 times in a set of execution traces. If it belongs to a set of temporal assertions that globally occur 1000 times in the same execution traces, the support of $A \to C$ is $10/1000 = 0.01$.

Definition 9. *Given a set of items I, and the corresponding set of transactions D, the correlation coefficient of the rule $X \to Y$ is the covariance of X and Y divided by the product of their individual standard deviations.*

More informally, the correlation coefficient can determine if antecedent and consequent are related or not by observing whether occurrences of the antecedent depend on occurrences of the consequent and vice versa. For example, Fig. 5 graphically shows the meaning of the correlation coefficient with respect to the association rule $X \to Y$. On the left, X and Y has a positive correlation, i.e., an increment in occurrences of X corresponds to an increment in occurrences of Y. In the middle, a negative correlation is shown. Finally, on the right, no dependence between X and Y exists. Higher is the correlation coefficient higher is the interestingness of the analysed rule.

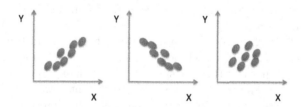

Fig. 5. The correlation coefficient: positive correlation (on the left), negative correlation (in the middle), no correlation (on the right).

6.2 Assertion Ranking

For estimating the interestingness of assertions, we implemented an assertion ranker based on support and correlation coefficient. The work flow of the proposed methodology is then divided in 3 main steps (Fig. 6):

1. *Counting of occurrences*: In this phase, the number of times an assertion is verified in the execution traces is computed. Then, each assertion is decomposed in antecedent and consequent and their respective frequencies in the execution traces are computed too.

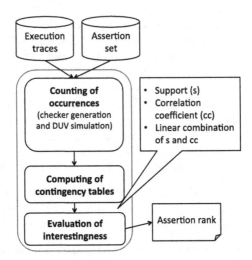

Fig. 6. Overview of qualification methodology.

2. *Computation of contingency tables*: the information collected in step 1 is then organized in contingency tables (one per each assertion) that represent the ingredients for the computation of the evaluation metrics in the final step. Contingency tables make simpler the extraction of information like how many times an antecedent and the corresponding consequent occur in the execution trace, how many times an antecedent occurs but the corresponding consequent does not, and how many times a consequent occurs but the corresponding antecedent does not.

3. *Evaluation of interestingness*: The final step, starting from the contingency tables, computes support, correlation coefficient, and their linear combination to obtain a final metrics that considers both of them. Their combination is necessary because support and correlation coefficient separately may provide very different estimations, which only partially characterise the quality of each assertion, as clarified later in this section.

In the following of this section the three steps of the proposed methodology are described.

Counting of Occurrences. To count occurrences of assertions, antecedents and consequents, we generate a checker for each assertion. A checker can be considered an automaton that monitors the evolution of the DUV during simulation and raises a failure when the corresponding assertion is violated [27]. To perform such a verification, the checker exactly knows when the antecedent is fired and when the consequent is then satisfied. Thus, it can be used for counting of occurrences as required for our estimation.

For example, the automaton generated for counting occurrences for an assertion like $always(p_a \rightarrow next(p_c))$ is reported in Fig. 7. The automata starts in the initial state *ant*. It remains in this state (corresponding to a vacuous satisfaction of the assertion) till the antecedent p_a is finally fired (transition $T3$). Then, it moves to the state *cons*, where it stays by continuously traversing $T4$ at each simulation instant while p_a remains true and p_c is also satisfied. This represents the case in which the assertion is activated and satisfied (non vacuously) for several consecutive simulation instants. The assertion is non vacuously satisfied also when the automaton exits *cons* by traversing $T7$, which corresponds to the case p_c still holds but p_a stops to be fired. Alternatively, the automaton exits *cons* to reach the *error* state through $T5$ in case p_c stops holding. In this case the assertion is falsified, but according to our assumption (assertions are true in the DUV) this never happens in our methodology. The number of occurrences of the assertion corresponds to the number of traversals of transitions $T4$ and $T7$. The number of times the antecedent is fired corresponds to the number of traversals of $T3$ and $T4$. Finally, the number of times the consequent is fired corresponds to the number of traversals of $T1$ and $T4$.

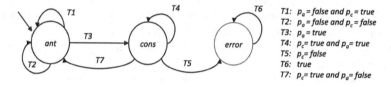

$T1$: p_a = false and p_c = true
$T2$: p_a = false and p_c = false
$T3$: p_a = true
$T4$: p_c = true and p_a = true
$T5$: p_c = false
$T6$: true
$T7$: p_c = true and p_a = false

Fig. 7. Example of the checker for assertion $always(p_a \rightarrow next(p_c))$.

Computation of Contingency Tables. Support and correlation coefficient can be effectively computed by relying on a 2×2 frequency count matrix called contingency table [28], whose computation derived from the counting of occurrences performed in the previous step. Given an assertion $A \rightarrow C$, its contingency table represents the relation between A and C. The cells of the table contain the following information (Table 1):

- Cell f_{11} is the number of times where A is true and consequently C is true in the execution traces;
- Cell f_{10} is the number of times where A is true but consequently C is false and other consequents than C are true in the execution traces, i.e., it is the sum of occurrences of assertions $A \rightarrow C'$ included in the considered assertion set with $C \neq C'$. It is worth noting that $A \rightarrow C$ and $A \rightarrow C'$ are not inconsistent, because C and C' refer to different temporal instants. For example, $always(p_1 \rightarrow next(p_2))$ and $always(p_1 \rightarrow next(next(p_3)))$ can be both true for the same DUV.
- Cell f_{01} is the dual of f_{10}, i.e., it is the number of times where A is false but A' different from A is true and consequently C is true in the execution traces, i.e., it is the sum of occurrences of assertions $A' \rightarrow C$ included in

the considered assertion set with $A \neq A'$. In this case, A and A' can also be conflicting because this doest not represent an inconsistency for the assertion set. For example, $always(p_1 \rightarrow next(p_2))$ and $always(p_3 \; until \; p_4 \rightarrow next(p_2))$ can be both true for the same DUV.

- Cell f_{00} is the number of times an assertion is true, whose antecedent and consequent are both different, respectively, from A and C, i.e., it is the sum of occurrences of the other assertions included in the analysed set.
- Cell f_{1X} is the sum of cells f_{11} and f_{10}.
- Cell f_{0X} is the sum of cells f_{01} and f_{00}.
- Cell f_{X1} is the sum of cells f_{11} and f_{01}.
- Cell f_{X0} is the sum of cells f_{10} and f_{00}.
- Cell f_{XX} is the grand total.

As an illustrative example, let us consider assertions reported in Table 2. For sake of clearness, and without loss of generality, the table does not show the atomic propositions composing antecedents and consequents of assertions, but only the temporal relations between them in PSL syntax [3]. The corresponding contingency tables are reported in Table 3. For example, for assertion $A1$, f_{11} correspond to the total number of occurrences of $A1$ in the analysed execution traces; f_{10} is equal to 0, since antecedent A does not appear in none of the other assertions; f_{01} is 0 since consequent $A \; until \; F$ does not appear in none of the other assertions; and finally, f_{00} is obtained by summing the occurrences of all the other assertions except $A1$. Cells f_{10} for assertions $A5$, $A6$ and $A7$ are not zero since they share the same antecedent E. Thus, f_{10} for $A5$, $A6$ and $A7$ are, respectively, the sum of occurrences of $A6$ and $A7$, $A5$ and $A7$, and $A5$ and $A6$. Similar considerations allow computing values for all the other cells of Table 3.

Table 1. Contingency table for $A \rightarrow C$.

	C	$\bar{C}$	
A	f_{11}	f_{10}	f_{1X}
$\bar{A}$	f_{01}	f_{00}	f_{0X}
	f_{X1}	f_{X0}	f_{XX}

Evaluation of Interestingness. Contingency tables provide basic ingredients for the computation of support and correlation coefficient of a temporal assertion. Concerning support, according to Definition 8, it is simply computed with the following formula:

$$s = \frac{f_{11}}{f_{XX}}. \tag{1}$$

The computation of the correlation coefficient for an assertion $A \rightarrow C$, according with Definition 9, is obtained instead by means of the following formula:

$$\rho = \frac{cov(A, C)}{\sigma A \cdot \sigma C} \tag{2}$$

Table 2. An assertion set with the corresponding number of occurrences in the execution traces.

Assertion ID	Assertion	Occurrence
A1	$always(A \rightarrow A\ until\ F)$	468
A2	$always(B \rightarrow B\ until\ G)$	436
A3	$always(C \rightarrow C\ until\ H)$	481
A4	$always(D \rightarrow D\ until\ I)$	361
A5	$always(E \rightarrow next(J))$	524
A6	$always(E \rightarrow next[2](J))$	516
A7	$always(E \rightarrow next[3](J))$	509

Table 3. Contingency tables of assertions reported in Table 2.

Assertion ID	f_{11}	f_{10}	f_{01}	f_{00}
A1	468	0	0	2827
A2	436	0	0	2859
A3	481	0	0	2814
A4	361	0	0	2934
A5	524	1025	0	1746
A6	516	1033	0	1746
A7	509	1040	0	1746

where $cov(A, C)$ is the covariance of A and C, while σA and σC are the standard deviation, respectively, of A and C. Disregarding mathematical steps, the correlation coefficient can be computed in terms of the cells of a contingency table as follows:

$$\rho = \frac{f_{11} \cdot f_{00} - f_{10} \cdot f_{01}}{\sqrt{f_{1X} \cdot f_{0X} \cdot f_{X1} \cdot f_{X0}}} \tag{3}$$

According to Eq. (1) the support ranks in the highest positions assertions that occur frequently in the execution traces. However, we can have very interesting assertions that occur a few times because they refer to corner cases. On the other hand, the correlation coefficient privileges assertions where the number of occurrences of the antecedent better matches the number of occurrences of the consequent, but assertions where these numbers are low could be extracted by chance without representing a real behaviour of the DUV. For this reason a combination of support and correlation coefficient provides a more accurate estimation of assertion interestingness. Thus, we propose the measure the interestingness of an assertion A through the following formula:

$$I(A) = \alpha * s_n(A) + (1 - \alpha) * \rho_n(A) \tag{4}$$

where, $\alpha \in [0, 1]$, and $s_n(A)$ and $\rho_n(A)$ are the value obtained by normalizing, respectively, the support s and the correlation coefficient ρ of A with respect to the whole set of analysed assertions. At varying of α the role of support becomes more or less important with respect to the role of the correlation coefficient in determining the final estimation of assertion interestingness. In our experiments best results have been obtained with $\alpha = 0.4$.

7   Experimental Results

Experimental results have been carried out on an Intel Xeon E5649 @2.53 Ghz equipped with 8 GB of RAM and running Linux OS. The benchmarks considered for evaluating the proposed mining strategy belong to the Open-Source-Test-Case (OSTC) platform developed as reference case study for the European project SMAC [29]. In particular, we considered the RTL implementation of the *UART* [30] and *BUS-APB* [31] components. These two benchmarks have been selected because they present different characteristics from the input/output latency point of view, i.e. the number of clock cycles required, at maximum, to compute the component's functionality. The I/O latency is an important parameter for mining approaches because longer is the I/O latency, higher is the time spent by the miner to create an assertion that puts in relation values provided to primary inputs with values obtained on primary outputs. *UART*, which is practically a parallel-to-serial/serial-to-parallel converter, requires 665 clock cycles before the output bit stream is produced, once data are provided in input for the conversion. On the contrary, the input/output latency of *BUS-APB* is 2 clock cycles.

Table 4 reports, for each component, the lines of code (*Lines*), the number of bits corresponding to control signals belonging to the primary inputs (*PIs*) and to the primary outputs (*POs*), and the input/output latency (*I/O latency*). Execution traces composed of 10,000 clock cycles have been generated for the two benchmarks by simulation.

The mining methodology proposed in this paper has been compared with a state-of-the-art approach presented in [5], which mines assertions from execution traces through an induction algorithm based on a decision tree [32]. The comparison between the two approaches is reported in Tables 5 and 6 concerning, respectively the characteristics of the mined assertions and mining execution times, and the quality of the mined assertions measured in terms of mutation coverage [14].

Columns 2 and 3 of Table 5 report the configuration parameters, i.e., the length of considered time windows (*tw_len*) (which corresponds to the I/O latency of the DUV), the maximum number of propositions allowed in the antecedent of the mined assertions (*max_len*) for the time window approach (i.e., the maximum number of clock cycles that are observed in the antecedent), and the maximum depth of the analysed decision tree (*max_depth*) for the approach proposed in [5]. The parameters *max_len* has been selected according to the characteristics of the DUVs. For example, *max_len* = 1 for *UART* because the values assigned to the input control signals to start the data elaboration are provided in a single clock cycle, while *max_len* = 2 for *BUS-APB* since input

Table 4. Characteristics of benchmarks.

DUV	Lines	PIs	POs	I/O latency
BUS-APB	390	6	12	2
UART	6819	10	9	665

control signals influence the bus functionality during the whole elaboration phase that always embraces 2 clock cycles. On the contrary, for the decision-tree based approach the maximum depth of analysed decision tree must be specified; we tested different values and we saw that for values higher than 10 and 12, respectively for *UART* and *BUS-APB*, the execution time of the algorithm increases without improving the quality (measured in terms of mutation coverage) of the mined assertions. Then, Columns 4–7 report the mining results, i.e., the number of extracted assertions (# *ass.*), the average number of input atomic propositions included in the antecedent of the extracted assertions (# *ant.*), the average number of output atomic propositions included in the consequent of the extracted assertions (# *cons.*), and the total time required for the mining procedure (*time*). Looking at the results, we see that the number of assertions generated by our approach is smaller than the number of assertions generated by [5]. However our assertions are composed of consequents with a higher number of atomic propositions, which reflects in a better description of the behaviours of the primary outputs of the DUV when an antecedent is fired. On the contrary, antecedents are generally compact (i.e., the number of involved atomic propositions is small), thus assertions cover a large number of behaviours from the perspective of the DUV's primary inputs. Finally, concerning execution time, our approach outperforms the decision tree-based algorithm when applied to mine assertions on DUVs, whose I/O latency (which impacts on the offset between antecedent and consequent) is very high, like in the case of *UART*.

Table 5. Number of assertions extraced by the time-window approach and the decision-tree based approach.

DUV	Configuration parameters		Results			
	tw_len	max_len	# ass.	# ant.	# cons.	Time
Time window-based approach						
BUS-APB	2	2	24	3.3	11.1	1 s.
UART	655	1	21	2.94	6.47	720 s.
DUV	Configuration parameters		Results			
	tw_len	max_depth	# ass.	# ant.	# cons.	Time
Decision tree-based approach [5]						
BUS-APB	2	12	86	2.82	1	1 s.
UART	665	10	39	5.6	1	5820 s.

Table 6. Comparison between the proposed approach and [5] based on mutant coverage.

DUV	# observ.	# covered	Avg	Time
Time window-based approach				
BUS-APB	22	22	10.27	70 s.
UART	149	99	26.85	4208 s.
DUV	# observ.	# covered	Avg	Time
Decision tree-based approach [5]				
BUS-APB	22	22	0.8	83 s
UART	149	58	9.08	46853 s

7.1   Assertion Qualification

Mined assertions by adopting our approach and the decision tree-based algorithm have been then compared to evaluate their effectiveness in covering the DUV behaviours. The comparison has been done in terms of *mutant coverage*, which represent small alterations of the DUV's source code that perturb its functionality. A mutant is observable if, in comparison with a mutant-free DUV, its effect is visible as an alteration in the DUV's primary outputs. A mutant is covered by an assertion if the assertion fails when the mutant is observed at primary outputs. The mutant coverage C is then the ratio between covered mutants and observable mutants. Uncovered mutants highlight the incompleteness of the assertions set [14]. The well-known bit coverage fault model have been selected to inject mutants in the control signals of the DUVs [33]. Bit coverage alters, in single fault mode, each bit of the affected signal by fixing its value to 0 (stuck-at 0) or to 1 (stuck-at 1). Table 6 reports the results of the mutation analysis by showing the number of observable mutants (# *observ.*), the number of covered mutants (# *covered*), the average number of mutants covered by each assertion (*avg*), and finally the time required to simulate the OSTC platform connected to the set of checkers[3] corresponding to the assertions mined for *UART* and *BUS-APB* in presence of one mutant (*Time*).

The mutant coverage achieved for *BUS-APB* is 100 % for both approaches, while the time window-based approach outperforms the decision-tree algorithm concerning mutant coverage of *UART*. Moreover, the number of mutants covered in average by each assertion mined with our approach is higher. Finally, concerning the simulation time, we observe that checkers corresponding to assertions mined by the decision tree algorithm require a longer simulation time, which greatly increases for assertions that predicate on DUV with a long I/O latency,

---

[3] A checker is an automaton that monitors the evolution of the DUV during simulation and raises a failure when the corresponding assertion is violated. We generated checkers for mined assertion by using IBM FoCs [34].

as in the case of *UART*. We observed in particular, that antecedents of assertions generated according to [5] are composed of atomic propositions that could be removed, since they do not affect the truth value of the assertions. This drawback is implicit in the use of a decision tree-based data structure, and it depends on the fact that an assertion generated at a leaf node necessarily includes atomic propositions predicating on variables involved in all previous levels of the tree. This leads to create assertions with longer antecedents, whose checkers require longer simulation times. Moreover, simulation times are affected by the total number of assertions which is higher in the case of the decision tree-based algorithm.

Finally, the accuracy of the interestingness estimation measured according to mutant coverage C has been compared with the metrics I we defined in Sect. 6. The hypothesis is that assertions with the highest mutant coverage are ranked in the highest positions also according to the proposed metrics I.

Table 7. Comparison between assertion ranking based on metrics I and mutant coverage C.

DUV	# ass.	# mut.	Preserved mutants				Loss mutants			
			$Q4_I$	$Q4_C$	$Q4_I \cup Q3_I$	$Q4_C \cup Q3_C$	$Q2_I \cup Q1_I$	$Q2_C \cup Q1_C$	$Q1_I$	$Q1_C$
UART	21	99	76	73	97	97	2	2	1	2
BUS-APB	24	22	18	NA	22	21	0	1	0	0

To experimentally prove the previous hypothesis, after the computation of the metrics I (with $\alpha = 0.4$) and the mutant coverage C, we divided assertions in 4 groups, respectively, $Q1_I, \ldots, Q4_I$ for I, and $Q1_C, \ldots, Q4_C$ for C. The division in groups has been done according with *quartiles* computed on I and C. In this way, the top 25%-ranked assertions with respect to I and C are included, respectively, in $Q4_I$ and $Q4_C$, while the worst 25%-ranked assertions are included in $Q1_I$ and $Q1_C$. Similarly, $Q3_I$ and $Q3_C$ include assertions between the first and the second quartile, while $Q2_I$ and $Q2_C$ include assertions between the second and the third quartile. Then, we analysed the impact of assertions belonging to the different groups in covering mutants. Results are reported, for assertions extracted according to our mining approach, in Table 7. After the DUV name, the second and the third Columns report, respectively, the number of analysed assertions (#*ass.*) and the number of mutants totally covered by assertions (#*mut.*). Then, Columns under *Preserved mutants* show how many mutants are still covered by preserving assertions belonging to only $Q4_I$ and only $Q4_C$, and to only $Q4_I \cup Q3_I$ and only $Q4_C \cup Q3_C$. Finally, Columns under *Loss mutants* show how many mutants remain uncovered by removing assertions belonging to $Q2_I \cup Q1_I$ and $Q2_C \cup Q1_C$, and to only $Q1_I$ and only $Q1_C$. It is evident from the results reported in Table 7 that measuring the interestingness of assertions according to the metrics I proposed in this paper ranks in the highest positions assertions that cover the most of mutants, while in the lowest positions remain assertions that very rarely cover mutants not yet covered by better ranked assertions. In this context, the ranking provided by I is even better than the ranking provided by C, since, for example, in the case of UART, 76 mutants are covered by assertions included in $Q4_I$, while only 73 mutants are covered by assertions included in $Q4_C$; on the opposite, only one mutant remains uncovered by discarding assertions in $Q1_I$, while 2 mutants remain uncovered by discarding assertions in $Q1_C$.

It is worth noting also that in the case of *BUS-APB*, the number of mutants covered only by assertions belonging to group $Q4_C$ cannot be computed, because due to a particular distribution of covered mutants among assertions, the third quartile correspond exactly to the fourth (i.e., to the maximum number of mutants covered by the assertions with the highest mutant coverage). In particular, this happens because, by chance, 8 assertions on 24 cover the same (highest) number of mutants. In this situation, due to the low variability of mutant coverage among assertions there is no distinction between $Q3_C$ and $Q4_C$. This represents a drawback of the mutant-based analysis, which is instead outcome by the approach proposed in this paper that can effectively distinguish between $Q3_I$ and $Q4_I$. A further analysis has been conducted by measuring the time required for the computation of I and C.

Results are reported in Table 8. It is evident that measuring I (I *time*) requires a few seconds, independently from the complexity of the DUV. On the contrary, mutation analysis requires a longer verification time I (C *time*) to simulate DUV and checkers for each mutant. This is particularly evident for complex designs like *UART*, where assertions predicate on large time windows (up to 665 clock cycles). For sake of clarity, the time reported for I does not include the time spent for counting assertion occurrences in the execution traces, since the result of such a counting is already available when assertions are automatically generated through assertions mining. If this information was not available, or assertions were manually defined, the time for computing I would include the time spent for one simulation run to compute assertion occurrences on the execution traces ($I + sim$ *time*), while computation of C always requires a number of simulation runs equal to the number of mutants.

Table 8. Execution time for computing I and C.

DUV	I time	I + sim time	C Time
UART	2 s	4208 s	26400 s
BUS-APB	2 s	70 s	940 s

From previous considerations we derive that the evaluation of the assertions according to the methodology proposed in Sect. 6 represents a faster and more effective approach for assertion qualification with respect to measuring the quality of assertions by using a mutant coverage-based approach.

8   Conclusions

The paper presents a mining algorithm for automatic generation of LTL temporal assertions and a qualification metric for the evaluation of the assertion interestingness.

On one side, the assertions generation technique relies on a time window-based analysis of execution traces that searches for behaviours that repeat periodically capturing the relation between primary inputs and primary outputs of the DUV. The approach is particularly suited for mining assertions that describe the behaviour of the control signals of the DUV, which are used to implement the I/O communication protocol surrounding the computation of the DUV core functionality. In comparison with a state-of-the-art assertion miner proposed in [5], experimental results show that our approach generates a more compact set of assertions, which achieves a higher mutant coverage and requires shorter times for the simulation of the corresponding checkers.

On the other hand, the qualification approach re-adapts metrics typically adopted in data mining, i.e., support and correlation coefficient, to measure the importance of an assertion on the basis of both its activation frequency during simulation runs and the correlation between its antecedent and consequent. Experimental results showed that, compared to traditional mutant coverage-based techniques, our metrics provides a better estimation of assertion interestingness by ranking in the top positions assertions that cover the major number of mutants and in the lowest positions assertions that cover mutants detected also by better ranked assertions. Finally, concerning estimation time, we outperform the mutant coverage-based approach of one order of magnitude, by considering also the time required for computing the frequency of assertions by simulation. When such frequencies are already available (e.g., when provided by an assertion mining tool) the computation of the proposed metrics is almost negligible (a few seconds).

References

1. Gupta, A.: Assertion-based verification turns the corner. IEEE Des. Test Comput. **19**(4), 131–132 (2002)
2. Pnueli, A.: Linear and branching structures in the semantics and logics of reactive systems. In: Brauer, W. (ed.) ICALP 1985. LNCS, vol. 194, pp. 15–32. Springer, Heidelberg (1985). doi:10.1007/BFb0015727
3. Standard for property specification language (PSL), IEC 62531: 2012(E) (IEEE Std 1850–2010), pp. 1–184 (2012)
4. Ammons, G., Bodík, R., Larus, J.R.: Mining specifications. ACM Sigplan Not. **37**(1), 4–16 (2002)
5. Hertz, S., Sheridan, D., Vasudevan, S.: Mining hardware assertions with guidance from static analysis. IEEE Trans. Comp. Aided Des. Integr. Cir. Syst. **32**(6), 952–965 (2013)
6. Danese, A., Ghasempouri, T., Pravadelli, G.: Automatic extraction of assertions from execution traces of behavioural models. In: proceedings of ACM/IEEE DATE (2015)

7. Jasper Activeprop. http://www.jasper-da.com
8. http://www.atrenta.com/solutions/bugscope.htm5
9. Bertasi, M., Di Guglielmo, G., Pravadelli, G.: Automatic generation of compact formal properties for effective error detection. In: Proceedings of ACM/IEEE CODES+ISSS, pp. 1–10 (2013)
10. Li, W., Forin, A., Seshia, S.A.: Scalable specification mining for verification and diagnosis. In: Proceedings of ACM/IEEE DAC (2010)
11. Katz, S., Grumberg, O., Geist, D.: "Have i written enough properties?" - a method of comparison between specification and implementation. In: Pierre, L., Kropf, T. (eds.) CHARME 1999. LNCS, vol. 1703, pp. 280–297. Springer, Heidelberg (1999)
12. Hoskote, H., Kam, T., Ho, P.H., Zao, X.: Coverage estimation for symbolic model checking. In: Proceedings of ACM/IEEE DAC, pp. 300–305 (1999)
13. Jayakumar, N., Purandare, M., Somenzi, F.: Dos and don'ts of CTL state coverage estimation. In: Proceedings of ACM/IEEE DAC, pp. 292–295 (2003)
14. Fedeli, A., Fummi, F., Pravadelli, G.: Properties incompleteness evaluation by functional verification. IEEE Trans. Comput. **56**(4), 528–544 (2007)
15. Tan, P.-N., Kumar, V., Srivastava, J.: Selecting the right interestingness measure for association patterns. In: Proceedings of ACM/SIGKDD KDD, pp. 32–41 (2002)
16. Tan, P.-N., Kumar, V.: Interestingness measures for association patterns: a perspective. In: Proceedings of Workshop on Postprocessing in Machine Learning and Data Mining (2000)
17. Lo, D., Maoz, S.: Specification mining of symbolic scenario-based models. In: Proceedings of ACM PASTE, pp. 29–35 (2008)
18. Lo, D., Khoo, S.-C., Liu, C.: Efficient mining of iterative patterns for software specification discovery. In: Proceedings of ACM KDD, pp. 460–469 (2007)
19. Henkel, J., Diwan, A.: Discovering algebraic specifications from java classes. In: Cardelli, L. (ed.) ECOOP 2003. LNCS, vol. 2743, pp. 431–456. Springer, Heidelberg (2003). doi:10.1007/978-3-540-45070-2_19
20. Ernst, M., Cockrell, J., Griswold, W., Notkin, D.: Dynamically discovering likely program invariants to support program evolution. IEEE Trans. Softw. Eng. **27**(2), 99–123 (2001)
21. Sheridan, D., Liu, L., Kim, H., Vasudevan, S.: A coverage guided mining approach for automatic generation of succinct assertions. In: Proceedings of IEEE VLSI Design, pp. 68–73 (2014)
22. Bonato, M., Di Guglielmo, G., Fujita, M., Fummi, F., Pravadelli, G.: Dynamic property mining for embedded software. In: Proceedings of ACM/IEEE CODES+ISSS, pp. 187–196 (2012)
23. Ernst, M.D., Perkins, J.H., Guo, P.J., McCamant, S., Pacheco, C., Tschantz, M.S., Xiao, C.: The Daikon system for dynamic detection of likely invariants. Sci. Comput. Program. **69**(1), 35–45 (2007)
24. http://www.atrenta.com/about-bugscope.htm5
25. Antunes, C.M., Oliveira, A.L.: Temporal data mining: an overview. In: Proceedings of Workshop on Temporal Data Mining (2001)
26. Bayardo Jr., R.J., Agrawal, R.: Mining the most interesting rules. In: Proceedings of the Fifth ACM SIGKDD International Conference on Knowledge Discovery and Data Mining, pp. 145–154. ACM (1999)
27. Boulé, M., Zilic, Z.: Generating Hardware Assertion Checkers: For Hardware Verification, Emulation Post-Fabrication Debugging and On-Line Monitoring. Springer, Netherlands (2008)

28. Pearson, K., Filon, L.N.G.: Mathematical contributions to the theory of evolution. IV. on the probable errors of frequency constants and on the influence of random selection on variation and correlation. Philos. Trans. **191**, 229–311 (1898)
29. http://www.fp7-smac.org
30. http://opencores.org/project,a_vhd_16550_uart
31. http://www.arm.com/products/system-ip/amba/amba-open-specifications.php
32. Quinlan, J.R.: Induction of decision trees. Mach. Learn. **1**(1), 81–106 (1986)
33. Fin, A., Fummi, F., Pravadelli, G.: Amleto: a multi-language environment for functional test generation. In: Proceedings of IEEE ITC, pp. 821–829 (2001)
34. https://www.research.ibm.com/haifa/projects/verification/focs/

Author Index

Printed in the United States
By Bookmasters